Library of
Davidson College

The Cultural Drama

The Cultural Drama
Modern Identities and Social Ferment

Foreword by S. Dillon Ripley

Edited by Wilton S. Dillon

Smithsonian Institution Press
City of Washington, 1974

Copyright © 1974 Smithsonian Institution

The essay appearing in this book under the title "Everyday Life and Social Identity" by Kenneth B. Clark was previously published in PATHOS OF POWER, copyright © 1974 by Kenneth B. Clark. Reprinted by permission of Harper & Row.

The Smithsonian Institution acknowledges with appreciation contributions in support of the Symposium by the staff and officers of the Charles F. Kettering Foundation and the Rockefeller Brothers Fund.

FRONTISPIECE: Indians watching the Crow Fair, Montana, 1941. Photo by Marion Post Wolcott from collection of Farm Security Administration. *Library of Congress.*

Smithsonian International Symposia Series

Knowledge Among Men, Paul H. Oehser, editor. Simon and Schuster, 1966.

The Fitness of Man's Environment. Smithsonian Institution Press, 1968.

Man and Beast: Comparative Social Behavior, J. F. Eisenberg and Wilton S. Dillon, editors. Smithsonian Institution Press, 1971.

PAPERS DELIVERED
AT THE SMITHSONIAN INSTITUTION SYMPOSIUM
NOVEMBER 16-19, 1970
Saul D. Alinsky
John W. Bennett
Peter L. Berger
Kenneth B. Clark
David Brion Davis
Vine Deloria, Jr.
Ivan Illich
Michio Nagai
Conor Cruise O'Brien
Alain Touraine

ESSAYS SUBSEQUENTLY CONTRIBUTED
TO THIS COLLECTION
Steven F. Arvizu
Wilton S. Dillon
Donald Keene
Murray L. Wax

Library of Congress Cataloging in Publication Data
Main entry under title:

The Cultural drama.

(Smithsonian international symposia series)
Consists chiefly of papers presented at a symposium sponsored by the Smithsonian Institution and held in Washington Nov. 16-19, 1970.
1. Assimilation (Sociology)—Addresses, essays, lectures. 2. Pluralism (Social sciences)—Addresses, essays, lectures. 3. Ethnic attitudes—Addresses, essays, lectures. I. Dillon, Wilton, ed. II. Smithsonian Institution. III. Series.
HN65.C83   301.24'1   74-6166
ISBN 0-87474-147-5

Smithsonian Institution Press Publication Number 5134

Designed by Stephen Kraft
Printed in the United States by Universal Lithographers, Inc.
Distributed in the United States and Canada by George Braziller, Inc.
Distributed throughout the rest of the world by Feffer and Simons, Inc.

First edition

TO Albert Gollin

Contents

13 FOREWORD *S. Dillon Ripley*

25 EDITOR'S NOTE & ACKNOWLEDGMENTS

PROLOG

33 E Pluribus Unum? *Wilton S. Dillon*

SCENARIOS

71 Actors, Roles, and Stages *Conor Cruise O'Brien*

DIVERSITY

89 The New Exodus *Vine Deloria, Jr.*

107 Cultural Pluralism, Political Power, and Ethnic Studies *Murray L. Wax*

123 Education for Constructive Marginality *Steven F. Arvizu*

IDENTITY

139 Cultural History and the American Identity *David Brion Davis*

159 Modern Identity: Crisis and Continuity *Peter L. Berger*

183 Everyday Life and Social Identity *Kenneth B. Clark*

197 Cultural Integrity and Personal Identity: The Communitarian Response *John W. Bennett*

237 Social Identity and the Formation of Social Movements *Alain Touraine*

PROTEST AND CHANGE

271 The Death of Mishima *Donald Keene*

289 The Double Revolution *Saul D. Alinsky*

305 Radical Alternatives to Schools *Ivan Illich*

EPILOG

321 Fear and Self-Transformation *Michio Nagai*

326 Suggested Further Reading

Foreword

S. Dillon Ripley
Secretary of the Smithsonian Institution

Biologist, ecologist, and authority on the birds of the Far East. Former Director of Yale University's Peabody Museum of Natural History. Many ornithological expeditions to the South Pacific, Ceylon, India, Nepal, and Bhutan. Former Fulbright Fellow, Guggenheim Fellow, and recipient of many honorary degrees. Author of **The Sacred Grove** *(1969), among numerous other publications.*

The peculiar glory of man is his ability to involve his fellows in a community, whether through the companionship of the hunt, the creation of symbols which become the basis of his institutions, the creation of works of art, or the establishment at Washington of an institution under the name "The Smithsonian Institution for the increase and diffusion of knowledge among men." With these latter words James Smithson bequeathed his fortune and set his enigmatic impress upon the future course of this institution. Here, in the midst of the massive temples of the world's most powerful democracy, we exist as a tiny enclave of encounter, a community of learning, a company of scholars accessible to everyone, a recipient of the nation's treasure, custodian of its past, and guardian, we venture to hope, of a legacy for its future.

This book reflects at least two aspects—temporal and continuous—of the life of this enclave of encounter and community of learning. One is the Smithsonian's practice of organizing international symposia as one of the many approaches to diffusing knowledge. Such symposia as the one that stimulated this book might be considered as another of the "short-lived phenomena" that are subjects of study of the Smithsonian Astrophysical Observatory in Cambridge, Massachusetts. Speakers and guests gather and go away. But unlike the circus, our tents stay up, and our own scientists, scholars, and others provide the continuities; they make up a noncontiguous community devoted to increasing knowledge about nature and man's place in it. The symposia serve as festivals punctuating the working year. The chapters of this book are mainly the work of our guests. They gathered November 16-19, 1970, as interpreters of diversity, identity, protest, and change in the modern world. They had been asked to examine how the present pace of change is manifest in cultural styles, and in the renewed human search for new identities, as old communities dissolve and their fragments coalesce in "new" forms. A Washington sociologist, Dr. Albert Gollin, described this particular symposium as "a happening on the Mall." But the temporal quality of a three-day encounter of minds must be seen against the historical backdrop of the Smithsonian and its more enduring concern with recording, for research, collections, and exhibitions, the lifestyles of past and present cultures and civilizations. Thus we have a better basis for guessing where we are going and even choosing where we want to go.

Architecturally and intellectually the Smithsonian mirrors styles and modes of other epochs. Yet it also reflects the recasting of older ideas in new forms or new contexts: a kind of cultural recycling, to borrow the ecologists' word. Nothing, however, is sure to remain the same as its turn comes up again.

Woodrow Wilson's espousal of the idea of self-determination applied to different historical circumstances in his day. Isn't it appropriate, though, to speculate on what it means for understanding what is going on inside many nation-states today? I refer to the strident demands for pride, autonomy, and recognition of various communities defined by ethnicity, race, sex, or age. The Woodrow Wilson International Center for Scholars now flourishes in its temporary home, the original headquarters building of our Institution. Our buildings and our ideas turn in purpose and use. The Smithsonian also once was the temporary home of the National Academy of Sciences, which moved on to develop a life of its own, its own institutional style, only to have its functions enlarged by Woodrow Wilson, who initiated the National Research Council. Now the Smithsonian shelters the Woodrow Wil-

S. Dillon Ripley

son Center in an altered physical setting as a memorial to the president-scholar, and his spirit is joined here by the common ancestor the Smithsonian shares with the Academy, the first Secretary of the Smithsonian, Professor Joseph Henry, that estimable prophet of America as a land of knowledge and inquiry. (America, no less than Confucianist China, animist Africa, or ancient Israel, needs its saints, prophets, and departed spirits who can be summoned from time to time for new periods of service.) While Professor Henry thought that the original Smithsonian building was too large for the Smithsonian, which he saw as a tiny nucleus of a national system of science and learning, he resolved that the Institution could in no wise be contained within its walls. Instead, it must reach forth as a kind of lighthouse for voyagers in their courses of discovery, a kind of beacon for men, to help them fulfill their destiny as part of a community of discoverers, a citizenry and enlightened public capable of collective determination of a course as a free people. And so he undertook to commit the Institution to an unheard-of goal—the popularization of knowledge for the nation at large. That original building was separated from the small cluster of important buildings in Washington by a broad, marshy meadow and canal, which he prevailed upon the Government of the District of Columbia to bridge so that citizens might come here to participate in discourse with the leading spirits of their age. In his original plan for the Institution, he wrote that essays should be commissioned from the foremost scholars and scientists of the day in an effort to close the gap between the learned community and men and women in everyday life. In our time as in his the spirit of democracy abhors a gulf between the citizen and specialist.

The Institution has initiated three previous symposia and books: *Knowledge Among Men, The Fitness of Man's Environment,* and *Man and Beast: Comparative Social Behavior.* These titles suggest that each was designed to illuminate some aspect of human concern, viewed also from whatever realm of expertise or understanding we in the Smithsonian could apply to it. We are trying to make the full round of scholarly specialties and combinations in the Institution so that different themes, though overlapping or seemingly repetitive, ultimately call upon the diversity of our resources. Thus our art, historical, ecological, biological, and behavioral scholars may be brought into play in one or more symposia while radio astronomers, oceanographers, and historians of space technology may have their special interests and those of the public reflected in another set.

Formerly, the Smithsonian published a series of commissioned or derived articles appearing in annual reports. These in themselves were descended from the earlier tradition of "Smithsonian Contributions to Knowledge," started by Professor Henry. The symposia also are an attempt to reproduce the older public lectures, which Henry started and which played such a vital part in the earlier cultural life of Washington. Professor Henry, himself a physicist, was much moved by human concerns.

In his address in 1874 at the laying of the cornerstone of the American Museum of Natural History in New York, he expressed social premonitions:

"Modern civilization tends to congregate the population of countries into large cities. . . . cities tend to increase more rapidly than the general population . . . [due to] the education of the working classes and the introduction of labor-saving machines. . . . cities in proportion to their extent and rapidity of growth engender habit of thought and of action of a character the reverse of progress and which, if unrestrained, would tend to disintegrate society and resolve it into its primitive barbarous elements . . . . these principles are eminently applicable in New York."

How right he was. In 1891 and subsequently Thomas G. Hodgkins gave $250,000 to the Smithsonian, stipulating that this fund should be used for studies of atmospheric air and the welfare of man. Like Henry, Hodgkins was oppressed by the cities and the medical-health problems of the poor living under a pall of coal smoke.

While physical and medical issues seemed to be at the fore insofar as man was concerned, research and collections at this Institution tended to evolve into two directions which have to do with man's cultural styles. On the one hand, man is a user of objects. Signally in America the nineteenth century brought to bear all Americans' inventiveness in the form of the struggle to develop machines to conquer the country and objects to improve everyday life. The Patent Office, when built in Washington, was the largest building of its time in the country. This was surely symbolic of the urge to get ahead with the job, which pervaded so much of the mood of Americans. Our collections are the result of everything from the Patent Office collections to the world's fairs.

I have always been fascinated with objects, for as I have said before, objects help to fashion culture just as culture fashions objects. Man has hardly evolved genetically since the Stone Age and is potentially just as susceptible to the influence of objects in learning today as he was in preliterate cultures. We are what we are partly because of the shape and feel of things around us. So the historical collections of the Smithsonian can, in effect, not only speak to us of what we were like in previous times, but their evolving styles can speak to us and, indeed, offer premonitions for the present and the future.

One of the articles published in the Smithsonian Annual Report of 1897 was by Havelock Ellis and was entitled "Mescal, A New Artificial Paradise." It was a reprint, to be sure, but it signified another facet of this Institution's continuing concern with society.

In establishing the Bureau of American Ethnology (BAE) in 1879, John Wesley Powell's emphasis was on the study of North American Indian languages. He emphasized the need to study and understand a people's language in order to understand their culture and history, and sponsored the recording and study of a multitude of Indian languages then widely spoken in North America. He also took a historical and conservationist viewpoint in sponsoring the recording of unique or less widely distributed languages that were in danger of being lost through disuse or change.

Frank H. Cushing, an ethnologist appointed by Smithsonian Secretary Joseph Henry, and later a member of the BAE, lived with and studied the Zuni Indians for five years (1880-1885). His extensive field work represented a longer period of time than the more celebrated residence of Bronislaw Malinowski among the Trobiand Islanders and their culture. Cushing's aim, like that of many early ethnologists, was to discover and record the old religious and cultural practices before they became further "corrupted" and changed by contact with non-Indian culture. He learned the Zuni priesthood and was felt by some Zuni to know more about their religious traditions than any one of them. He was a pioneer of modern participant-observer anthropology.

James Mooney, a member of the BAE for thirty-six years, spent his lifetime in studies of the eastern Cherokee and the Indians of the Plains—Cheyenne, Arapaho, and Kiowa. He pioneered in studies of the Plains' Sun Dance and Ghost Dance; and Indian use of mescal and peyote. At the World's Columbian Exposition in Chicago in 1893 he was responsible for many of the exhibits showing Indian culture, and at the Trans-Mississippi and International Exposition in Omaha in 1898 he brought Indian family groups to the fair to construct their typical dwellings and to demonstrate native crafts. In a newspaper interview in 1892 he was quoted as saying of the Indians, "They like me because I come to them in sympathy, eager to preserve all that is sacred to them while the missionary and the agent come to do away and destroy their traditions."

William N. Fenton, a member of the BAE from 1939 to 1952 (later Director of the New York State Museum at Albany and currently Research Professor of Anthropology at the State University of New York at Albany), has concentrated on the study of cultural continuity and change in Iroquois culture. Elaborating and refining approaches used by Frank Speck of the University of Pennsylvania and John R. Swanton (a member of the BAE from 1900 to 1944), he has combined the utilization of historical documents with the techniques of modern anthropological field work in an ethnohistorical approach called "upstreaming"—proceeding from the known to the unknown, and from recent sources to earlier ones.

William C. Sturtevant, BAE member, 1956-1965, now our curator of North American Ethnology, has used similar techniques in studies of Seminole culture, and has added the analysis of quantities of dated photographs as a method of documenting and studying cultural change. His work reflects, in part, the pioneering theoretical work of the late Alfred Kroeber of the University of California on cultural styles and their configuration.

The Institute of Social Anthropology was established under the directorship of the late Julian H. Steward on September 8, 1943, during wartime, as an autonomous unit of the Bureau of American Ethnology. Its purpose was to set up cooperative institutes of social anthropology for scientific research and training in certain Latin-American countries, each working in cooperation with the Smithsonian Institution. This program was carried out under the auspices of the interdepartmental Committee on Scientific and Cultural Cooperation in Latin America of the U.S. Department of State, and was financed by funds transferred from the State Department to the Smithsonian Institution.

Dr. Steward, later of the University of Illinois, directed the Institute of Social Anthropology until September 1946, when he was succeeded by George M. Foster of the University of California, Berkeley, who continued as Director until the Institute was abolished in 1952.

At its outset, the Institute staff consisted of eight social scientists (four social anthropologists, two cultural geographers, a linguist, and a sociologist), stationed in Brazil, Colombia, Mexico, and Peru (for a short time in 1950 a station was established in Guatemala, but it was discontinued the same year for lack of funds). Their duties consisted in instructing local students (in collaboration with Latin-American universities and other institutions) in social science research techniques through classroom, laboratory, and field situations, and in assisting with publication projects.

The Institute continued in this capacity until 1949 when it was placed under the Division of International Exchange of Persons, another committee of the Department of State. The Institute did not form an organic part of the new program, and the Department of State decided to terminate its support as of December 31, 1951.

So much for that effort. It was somewhat of a pioneer one for the Smithsonian; it also signaled a change in direction for anthropology. Social anthropology has continued ever since, though showing variance from the British school. The Smithsonian continues to experiment with various approaches to anthropology. The traditional tasks of research and curatorial activities in archeology, physical anthropology, and ethnology are carried on by the Department of Anthropology of the National Museum of Natural History alongside newer developments by the Center for the Study of Man: long-range work on *A Handbook of American Indians,* as well as a kind of "urgent anthropology" having to do with recordings, linguistic studies, and documentations of the remaining tribal and subcultural identities before they are submerged and homogenized completely. Studies also are under way on differential human fertility, on man's perception of his environment, and on education as cultural transmission. These latter initiatives are an important Smithsonian link with contemporary social and behavioral sciences.

From here, where does anthropology go? Everybody I ask gives me a different answer. I do not believe it should become an antiquarian study any more than the historical-technology and scientific collections imply that our Museum of History and Technology is populated by antiquarian curators. In fact, we have a Division of Performing Arts proving the contrary. Our Folk Festival, held each summer, in effect brings the instruments and the tools out onto the Mall to prove to people that mankind somewhere in America still knows how to shear sheep, card wool, weave cloth, pickle vegetables, and make and play a dulcimer or a violin. And Dr. Richard Ahlborn, our curator of ethnic history, has provided Texas educational TV outlets with photographs of objects suggesting the past and current contributions of Spanish-speaking people to American life. We applaud such efforts as the Grand Rapids Public Museum to stage community festivals as expressions of our debts to cultural styles brought by Mexican immigrants. Our own Anacostia Neighborhood Museum has focused on African underpinnings of

American civilization. Down with homogenization, I say! There are millions of Americans today who will say "Amen" to that.

Even though sometimes charged with being a sacred cow, the Smithsonian is here to learn as well as teach. We welcome the juxtaposition of a living tableau of ladies from Women's Lib on a Smithsonian stage against a poster advertising our exhibition of nineteenth-century art, "The Genteel Female." We enthusiastically concur that sisterhood is powerful and that men, too, will be liberated by such power. We need more of the iconoclastic wit of women who placed an apple on the Smithsonian mace to question our "authority" as an establishment conduit of past cultures. As students have suggested, we need to experiment with the format of our symposia so that we roll with the modern punches and enjoy the participatory democracy of discussion by others than pedigreed guest scholars past the age of thirty. The possibility of myopia as an endemic condition in our nation's capital prompts us to pay close attention to protest and change as an essential part of contemporary cultural history. We also accept change for ourselves, and hope our setting will not be construed as the relics of old dead things but rather the reminders of the constant thread of continuity. Yet we do not wish our enclave to make us immune from encounters with reality. Past is present. What we speak of is already past. The "now" generation gives constant reminders that no one has ever experienced before what is happening just now. But even as we write these words we are all a little older, and this too will pass.

I have a conviction that young people today are more capable of learning in relatively unstructured situations, just as the new media are making us once again remember our preliterate gifts of ear and hand and nose and the nonreading eye. If this is so, then museums are part of the educational turf of the future. Open education is akin to the maxims of those sensible people who have always known that the only education is self-education achieved through discipline and motivation. So many universities today are agonizing through the welter of indecision created by the conscious craving of those young who wish to be educated, contrasted with the others who seek only training and the transfer of information and the acquisition of skills.

Open education is nearer the first, for it implies the conscious will to learn. But our formal state-approved educational apparatus has long since abandoned such goals in its desire to mold and conform. I sometimes think self-education is the bane of educators. No wonder, then, that the young student is often confused and disheartened.

The Carnegie Commission report on higher education and mental health may well presage a trend in the right direction, provided we can move—perhaps with the help of Ivan Illich—the juggernaut of education that exists throughout this land. Let us in the process encourage diversity and self-education, and look for emerging styles and fresh interpretations of change in a positive and encouraging way. For the young, I would say there is only one solution—love life, and in so doing be interested, for there is all in all in that, and all you will ever have.

One young writer and photographer, Richard Balzer, a Philadelphian, honored his generation and our Institution with spontaneous words delivered at the Smithsonian. While he observed that not one of the contributors to this volume sympathized with insights that allow little or no hope—insights that allow little or no faith—I continue to believe that he shares with some of us elders that love of life manifest in changing styles and forms of which this book is one small celebration. He brought us the wisdom of the historically minded and empathetic young by his assertion: "There are traditional radical ways. Traditional does not mean not being radical."

S. Dillon Ripley

Editor's Note & Acknowledgments

When the idea of a theater is inadequate or lacking, we are reduced to speculating about the plight of the whole culture.... Unless the cultural components of our melting-pot are recognized, evaluated, and understood in some sort of relationship—our religious, racial, and regional traditions, and our actual habits of mind derived from applied science and practical politics, seen as mutually relevant—how can we hope for a medium of communication more significant than that of our movie-palaces, induction centers, and camps for displaced persons? The ultimate questions about the theater of human life in our time, and the drama of the modern world, are interdependent, theoretical and practical at once; and therefore unanswerable.

FRANCIS FERGUSSON

The Citizens of America ... are, from this period, to be considered as the Actors on a most conspicuous Theatre, which seems to be peculiarly designed by Providence for the display of human greatness and felicity....

GEORGE WASHINGTON

"Man and Beast: Comparative Social Behavior," the Smithsonian's third international symposium, deliberately ignored race, ethnicity, and culture as differentials in human behavior. We concentrated more on how scientists and philosophers might extrapolate from animal behavior to understand ourselves as part of nature. In the fourth international symposium, the emphasis shifted to the mysterious processes by which human cultures change and continue to unfold in infinite variety while humans biologically remain a single species.[1]

Protest, so typical of the explosive social turmoil in American life in the late sixties, took quite different forms compared with changes set into motion by the American Revolution (to which special attention will be given in various symposia marking the Bicentennial celebrations in 1976). Yet, interpretations of protest and diversity, written in the middle of the verbal battles and confrontations of the period, remain useful points of departure for understanding our 200-year history and its various climaxes, as individuals and groups compete, fight, or cooperate in their quest for a livelihood and meaning in life.

Despite the richness of perspectives in this volume, I regret that H. L. Mencken and Mark Twain were not around for the symposium. Their evocative language, such as "whooping and clapper-clawing" and "tears and flapdoodle," would have matched the "Helzapoppin" atmosphere of the gathering of scholars and social critics. Tom Wolfe did show up, however, fresh from writing "Radical Chic: That Party at Lenny's,"[2] and dazzled the audience with a free-associational flow of words interspersed with electronic images of popular culture, which unfortunately could not be reproduced here.

The Cultural Drama—as the title of this volume—grew largely out of the Smithsonian's having become living theater during the symposium, which was entitled "Cultural Styles and Social Identities: Interpretations of Protest and Change." Onto the stage of the National Museum of Natural History came actors demanding time to be heard. Protest was to be seen, heard, experienced—not merely interpreted. "Data" were served up raw. Domingo (Nick) Reyes, founder and president of the National Mexican-American Anti-Defamation Committee, Inc., and Armando R. Rendon, Chicano author, gave new significance to Pirandello's drama *Six Characters in Search of an Author*. Their unrehearsed interruption of the opening ceremony brought into the Smithsonian a style then already known to universities, churches, and corporations: the verbal assault known as "confrontation." From a seat near the front came another action not in the original script, the cry of "hypocrisy," from Gabrielle Edgcomb, art historian and poet, protesting the absence of women in the formal program. In retrospect, their actions gave common meaning to the theme of the symposium, and literally dramatized the truths of Conor Cruise O'Brien's paper, "Actors, Roles, and Stages," which was read first. In his earlier book, *The United Nations: Sacred Drama*, Cruise O'Brien observed: "Almost all action . . . involves an element of acting, in the theatrical sense; sociologists remind us of the number of roles we play in our daily domestic, social, and professional lives. National politics also, whether democratic or authoritarian in form, has always required play-acting, symbolism, and ritual."[3] How true ring Cruise O'Brien's words when one reflects on Alcatraz, Wounded Knee, the return of the POW's, Watergate, or interprets the meaning of Mishima's suicide, when considering symbolic protest in a culture other than our own. How true they rang when read at the Smithsonian, situated on a Mall which, in recent years, has become increasingly used as a stage for dramas of protest. If change does not always follow protest, at least the actors who camped, sang, marched, and spoke along the Mall and around the monuments, whether hawks or doves, provided themselves with self-respect for having done what they regarded as a civic duty, acting out ritually their values or anger, and showing their willingness to weep from live tear gas, the stage prop of that epoch. Robert Brustein, Dean of the School of Drama at Yale University, aptly has described politics, Washington's industry, as theater.

Thus, to those who added much to the liveliness of the discourse, including bright young students who protested the absence of youth from the forum, we acknowledge a debt. They are now an essential part of our understanding of the tendency of the times to demand ethnic, age, or sex representation, while each group or category reserves the right to reject self-appointed or invited spokesmen. Nobody is "representative." The contributors to this volume—those who were commissioned to write papers and those whose essays were invited later—give new insights into how culture works as an organized group of learned responses; how all cultures, even the simplest, are marked by continuous change; and how the events which produce change, or slow it down, are so structured that they resemble a drama or a play. The essayists serve as a kind of Greek chorus brought to the Smithsonian from academia and other centers of thought and action.

Dr. Michio Nagai, the distinguished Japanese sociologist and editor, who served as chairman of the symposium, participated unwittingly in a drama which unfolded in the wake of the symposium. After addressing the final banquet, Professor Nagai left Washington for New York to visit his friend, Prof. Donald Keene, historian of Japanese literature. He discovered upon arrival that Keene had just been informed by Tokyo telephone of the suicide of Mishima, the Japanese novelist, in full view of press and television. Mishima's novel *After the Banquet* had not anticipated such a coincidence, but suggests now the enduring strength of literary metaphor if followed by tragic action. Keene's later interpretation of the fabled suicide, presented at the Smithsonian's Freer Gallery, is included in this volume.

In addition to the "extras" in the symposium, the regular cast identified with the effort included Philip C. Ritterbush, who conceived the original topic, David Chase, Peter Jessen, Dorothy Richardson, Maureen Flynn, Frances Hays Miller, Joel Shimberg, and William Wing. Consultants whose ideas were sought, but who are not responsible for the final form include Meredith Johnson, Julian Euell, Malcolm Watkins, Margaret Mead, Sam Stanley, William Sturtevant, Peter Marzio, Albert Meisel, and Irving Zaretsky.

The staff and officers of the Charles Kettering Foundation and the Rockefeller Brothers Fund, moreover, contributed valuable help above and beyond the call of financial support. And I should be remiss if I did not cite the many contributions to the book's finished form by Hope Pantell, Smithsonian Institution Press.

If another symposium were to be held today, different illustrations of "cultural drama" might well emerge. The Six-Day War mentioned by Conor Cruise O'Brien, for example, could be augmented by references to the Yom Kippur War. I have chosen, however, to present these interpretations as they were conceived (with only the most basic updating), rather than try to make them compete with the evening television news for contemporaneity.

W.S.D.

NOTES

[1] Four basic questions should be asked in summing up knowledge of culture change: (1) What are the internal and external factors that generate shifts in rates and types of culture change? (2) What are the processes by which culture change takes place? (3) What models and methods are now available for the study of culture change? and (4) How is the concept of culture change related to the closely associated phenomena of diffusion, innovation, evolution, acculturation, and nativism? See EVON Z. VOGT, "Culture Change," in *International Encyclopedia of the Social Sciences,* edited by D. L. SILLS (New York: Macmillan, 1968).

[2] *New York Magazine,* volume 3, number 23, June 8, 1970.

[3] See CONOR CRUISE O'BRIEN and FELIKS TOPOLSKI, *United Nations: Sacred Drama* (New York: Simon and Schuster, 1968).

Prolog

E Pluribus Unum?
Wilton S. Dillon
Smithsonian Institution

Anthropologist and educator. Director of the Smithsonian's Office of Seminars and international symposia series since 1969. Coeditor, with John F. Eisenberg, Man and Beast: Comparative Social Behavior *(1971), proceedings of the third Smithsonian symposium. Adjunct professor of anthropology, University of Alabama, since 1971. President, board of directors, Institute of Intercultural Studies, American Museum of Natural History, New York, and Trustee, Phelps-Stokes Fund of New York. Author of* Gifts and Nations *(1968) and contributor to various journals.*

Cultural diversity and heterogeneity counteract the tendency to cultural entropy. DAVID BIDNEY

Entropy is the general trend of the universe toward death and disorder. J. R. NEWMAN

This is not a book about the drama of a great American prophet, but the life and death of Martin Luther King, Jr. (1929-1968) provide vital substance for understanding the interplay between the concepts around which this book is organized. Think of King's incredible career—from Atlanta to Boston, Montgomery, St. Augustine, and Selma to Washington, D.C., and Stockholm to Memphis—when reading Conor Cruise O'Brien's essay on "Actors, Roles, and Stages."

The diverse elements in King's symbolic behavior—aimed at improving the civil rights and potentialities of all Americans—came out of the cultures of ancient Israel, Greece, and Rome (the Judeo-Christian tradition), and the more recent cultures of Africa (his slave ancestors and their cosmologies), India (Gandhi), New England (Thoreau), and Germany (Martin Luther's reformation). Such diverse sources of his role and identity as a spiritual leader, also black, give the lie to any simple classification of human beings based only on ethnicity or race. Essays in this book by David Brion Davis, Peter L. Berger, Kenneth B. Clark, John William Bennett, and Alain Touraine are useful points of departure for thinking of the processes shaping the formation of King's identities and how they, in turn, prompted his protests and the social movement capable of changing social structures and values perhaps more profoundly than, say, Mishima's suicide (interpreted by Donald Keene) has yet affected Japan's industrial civilization.

What Michio Nagai wrote about "Fear and Self-Transformation," moreover, provides an interesting counterpoint for King's personal fearlessness, either of jail or death. His martyrdom by those whose "counterprotest" (fear of King's successes?) produced an unforgettable televised dramaturgy around his Ebenezer Baptist tomb in Atlanta provides the raw material out of which modern educators, taking their cues from Ivan Illich, can fashion a vast new moral and civic curriculum without depending on schools. The King patrimony, like Watergate, is a system of education. In the concrete is the whole. One of the many lessons to be learned from King, and argued about, concerns the paradox of human separateness and belonging. Are not we all benefiting from King's marginality? How should society be organized if we were starting all over again? What fresh approaches are there to understanding *e pluribus unum*? What follows are some efforts to find these approaches.

Arguments for Diversity

"Down with homogenization!"

In his foregoing perspective on the historical stake of the Smithsonian Institution in opting for cultural pluralism as a vital feature of the American way of life, Secretary S. Dillon Ripley thus states smartly his preference for what should be avoided in the worldwide cultural revolution in which the peoples of the United States are caught up as we approach our third century as an organized society. If for no other reason than to prevent boredom among twenty-first-century museumgoers looking at artifacts from the material culture of this epoch, the Smithsonian has a bias in favor of diversity. Such diversity can be generated and maintained by various forms of protest, another theme of this book.

Up with what, if down with homogenization? Constructive marginality is one of the suggestions offered in this book by a Chicano anthropologist, Steven Arvizu. Diversity by design is the recommendation of Samuel B. Gould and

Dr. Martin Luther King, Jr., addressing the historic August 28, 1963, demonstration for civil rights at the Lincoln Memorial in Washington, D. C. Courtesy of Ebony Magazine.

his colleagues in a new book by the same title reporting on the work of the Commission on Non-Traditional Study. Nathan Glazer and Daniel P. Moynihan a decade ago began to look beyond the melting pot in New York City as a harbinger of things to come elsewhere in the United States. "Religion and race define the next stage in the evolution of the American peoples. But the American nationality is still forming: its processes are mysterious, and the final form, if there is ever to be a final form, is as yet unknown."[1]

Under the auspices of the American Academy of Arts and Sciences, with support of the Ford Foundation, Glazer and Moynihan are broadening their study of ethnicity in our time to look at ethnic problems from a worldwide perspective, and to examine the importance of ethnicity for social policy and social science theory.

In his essay, "The New Exodus," in this book Vine Deloria, Jr., a Standing Rock Sioux lawyer of scholar-churchman, warrior-chief ancestry, writes "The Constitution, as it has come down to us through two centuries of hardship and pain, can be the tested ground rules of the redefinition of society according to the uniqueness and integrity of our respective constituent groups. In this sense it shows every promise of being comparable to the laws of Moses, delivered in the wilderness to the tribes of Israel. . . . In that context the smallest tribe was equal to the largest and most powerful." This echoes the Rev. Andrew Young, a black member of Congress from Georgia, who, in a New Year's Day TV talk show, called for all Americans to understand the notion of "cultural equity."

In the world of museums, we have begun to hear phrases like "cultural democracy" alongside "cultural equity." At the 1972 meetings in Mexico City of the American Museums Association, a session was devoted to "A Curator for the Future," urging future curators to link past, present, and future by adherence to these propositions: "A cultural democracy exists where the culture of the nondominant elements in a society are accepted along with those of the dominant. A cultural democracy provides for an acceptance of historical, ethnic, and racial identity just as a political democracy represents the individual."

A new academic journal, *Ethnicity*, made its appearance in 1974. And a sampling of newspapers, magazines, journals of sociology, politics, law, anthropology, and publishers' book lists also shows a great preoccupation with ethnicity. For example, in April 1972 Alfred Kazin, the critic, and James T. Farrell, the novelist, told City College students in New York that the ethnic identity of minorities—which appeared to them in the 1930s as powerlessness—had contributed more creative power to American life and literature than any dream of a melting pot. "The melting pot was essentially an Anglo-Saxon effort to rub out the past of others and turn Europe into a place where nobody spoke English," Farrell said. Kazin said he and others in the City College class of 1935 were hyphenated people with an exaggerated respect for America, meaning Anglo-Saxon culture. In the 1940s the United States began to produce Jewish writers with a short usable tradition in America—Saul Bellow, Bernard Malamud, Philip Roth. Such other writers as Richard Wright and James Baldwin helped to make ethnic experience the basis for some of the richest, most interesting and creative elements in American literature. And two American cultural analysts of still other backgrounds have recently addressed themselves to the rise of ethnicity and ethnic awareness as striking features of modern American civilization, despite our heroic claims to assimilating the alien. Andrew Greeley, the Irish priest-sociologist-educator of Chicago, recently enlarged the concept of ethnicity to include multi-ethnic intellectuals who behave like any other ethnic group;[2] and Michael Novak, the philosopher now at the Rockefeller Foundation, in his book *The Rise of the Unmeltable Ethnics*[3] asks, among other things, why doesn't a curator who will organize an exhibition of black art organize one on Lithuanian art?

All of these proclamations and writings are part of a healthy and hopeful search for an answer to the question: If down with homogenization and the melting-pot theory, what do we Americans put in place of the old self-images and guidelines? How much of one's self is defined by one's citizenship? What modern meanings can be attached to *e pluribus unum*, an unexamined concept appearing on our coinage but absent from public discourse? (These questions hopefully will find some answers out of new federal- and foundation-supported research such as the National Advisory Council on Ethnic Heritage Studies of the Department of Health, Education, and Welfare, and the Afro-American, Mexican-American, and American Indian fellowships of the National Endowment for the Humanities.)

In the musical *1776* the incipient American "national

character" could be seen forming along heterogeneous lines. Though ethnicity as a concept was not in fashion, urban, rural, and regional differences were acted—or danced—out through such personalities as Adams of Boston, Franklin of Philadelphia, and Calhoun of a South Carolina plantation. Even Mr. Jefferson's distractions while drafting the Declaration of Independence could be regarded as a forerunner to the contemporary bumper sticker claim to Old Dominion distinctiveness: "Virginia is for lovers." A hundred years after the birth of the nation, Henry Adams was educating himself to these regional differences as the United States celebrated its Centennial.

The most repressive totalitarian cultural systems, taut with the sanctions of police, propaganda, and ideological arbiters, seem unfree of human needs to nurture diversity in some politically palatable form. Humans persist in finding—and celebrating—the uniqueness of persons or social categories whether the state or other big organizations (universities, corporations) like it or not. On the threshold of our 200th anniversary as a republic, the United States still is in search of a philosophy or a rationale for that web of human behavior we abstractly call "diversity," "dissent," "deviation," or "pluralism," and is seeking how to reconcile that search with the political demands of "solidarity," "loyalty,"

> The origin of e pluribus unum, the motto adopted by our Founding Fathers, is obscure. It may well have been considered a phrase too familiar to require any explanation of derivation. By the time of the Revolution, it had appeared for many years—together with the device of a hand grasping a bouquet—on the title page of The Gentleman's Magazine, a popular periodical in the Colonies. One nineteenth-century writer traced almost the same phrase (e pluribus unus) to a Latin poem, "Moretum," attributed to Virgil. The moretum was a type of thick soup or stew made of various ingredients that were ground up with a pestle. (See History of the Flag of the United States of America by Geo. Henry Preble. Boston: A. Williams and Company, 1880, pages 694-697.)

Wilton S. Dillon

THE Gentleman's Magazine:
OR,
Monthly Intelligencer.

For the YEAR 1731.

CONTAINING,

I. ESSAYS *Controversial, Humorous,* and *Satirical*; *Religious, Moral,* and *Political*: Collected chiefly from the *Public Papers.*

II. Select Pieces of POETRY.

III. A succinct Account of the most *remarkable Transactions*

and *Events* Foreign and Domestick.

IV. *Births, Marriages, Deaths, Promotions,* and *Bankrupts.*

V. The Prices of *Goods* and *Stocks*, and Bill of *Mortality.*

VI. A Register of Books.

VII. Observations in *Gardening.*

With proper INDEXES.

By SYLVANUS URBAN, Gent.

VOL. I.

Prodesse & delectare.   *E Pluribus Unum.*

LONDON
Printed, and sold at St *John's* Gate, by F. *Jefferies* in *Ludgate-street*, and most Booksellers.

E Pluribus Unum?

Library of Davidson College

and "security" as perceived by persons temporarily in power. The essays of this book revolve mainly around the interdependent themes of diversity, identity, protest, and change. They are highly pertinent to the search for a framework, other than the melting pot, in which we can think about, and perhaps enhance, the quality of life in the beginning of the third century of the American experience.

In "The Double Revolution," the late Saul D. Alinsky, reflecting his spirited protest to the end, seemed to be anticipating in 1970 the pertinence of his remarks for the national inventory called the Bicentennial: "Suffice to say that if the Founding Fathers were writing the Declaration of Independence today it would read: 'We hold these truths to be relative and self-evident,' for it is today undeniably self-evident that all truths are relative."

Cultural relativity—Ruth Benedict's and other anthropologists' analog to Poincaré's and Einstein's relativity, Heisenberg's indeterminacy, and Bohr's complementarity—may come to be a paradoxical absolute: living in today's complex world may require acceptance of relativism as a fixed anchor around which to build our lives, a kind of moving equilibrium in which tolerance of ambiguity is the condition for survival.

Benedict's collaborator, Margaret Mead, spoke less pompously to the modern understanding of the Declaration of Independence when she, comparing individuals with nations, remarked that the central message sent to the British in that eloquent eighteenth-century document was: "Look, I am now grown up; go to hell."[4] In the same speech she said that it is "very American" to incorporate those who protest or demand change by "putting them on the board of trustees," a less bloody way of carrying on a revolution or effecting reform. The British were relegated to the rank of honorable ancestors rather than governors-cum-parents, and then we proceeded to try to grow up by combining our unique New World experience with more eclectic models brought in by non-Anglo immigrants. That growing up process is still going on. Adulthood and maturity are "relative." Growth and learning and curiosity ideally should continue

Wilton S. Dillon

until death in the life cycles of individuals or nations. Does that mean, I wonder, that our chosen, rather than assigned or hereditary, governors or rulers today will follow that classic American pattern by incorporating the cries of the ethnic protestants into some new political rhetoric which might say, "Be American, Be Different!" or "Celebrate Diversity!" Would such slogans give us new content for the Jeffersonian phrase "the pursuit of happiness"? Could happiness become liberated as a concept from current notions of hedonism and material living standards and, instead, include the idea that "a sense of well-being" is what we mean now by happiness, and well-being includes the pleasure of searching for one or more ethnic identities as a legitimate leisure-time activity of fully employed citizens?

For the Smithsonian, at least, Secretary Ripley already has pointed toward a palatable public and personal partial answer to the question, "If down with homogenization, up with what?"

"We are a conservation organization," he wrote about the Smithsonian in a preface to the 1973 Festival of American Folklife program, "and it seems to us that conservation extends to human cultural practices." He described the Folklife Festival as a valuable asset for our role as preserver and conservator of living cultural forms, "not a vaudeville show," but rather "a demonstration of the vitality of those cultural roots which surround us and are so often overlooked," those which fall in the shadow of official civilization.

"The fact that we celebrate ethnic diversity in our culture is, I think, extremely important. We have too often thought of the Bertrand Lindsay-like concept of the United States being 'the Great Melting Pot,' the great homogenizing element in Western Civilization. But, as we've discovered, this is by no means true. It is worthwhile being proud, not fiercely proud, but gently and happily proud of the continuance of those cultural roots and their observances and practices which we celebrate. This is a festival that celebrates people who celebrate themselves—people who know who they are and where they came from."

If ethnicity is merely a hobby, a pursuit appropriate to

one's leisure time—going home to comfortable foods, jokes, and habits after being in the national mainstream which flows through school, office, or factory—an examination of what goes into the mass media today will serve up indicators that the identity question is far more pervasive than even Archie Bunker. Whether leisure-time activity or not, the search for identity, the rediscovery of lost continuities with cultures out of which ancestors came to the New World, constitutes a drama of profound interest to a whole range of Americans and American institutions: the politicians appealing to the "ethnic vote," the entertainment industry, the museums pressured to give "equal time" to

Wilton S. Dillon

minorities on their boards and in their exhibition cases, the social work and psychiatric professions,[5] labor unions, philanthropic foundations, and even overseas corporations which supply goods (Japanese-made Kachina dolls, for example) to the American ethnic market. The economics of ethnicity is increasingly important, but unstudied.

"Tribute to Tamburashi" was one of the themes of the Smithsonian's 1973 Festival of American Folklife, bringing together participants from Yugoslavia and their cultural descendants in America in demonstrations of songs and dances, customs and lore.

E Pluribus Unum?

The *Wall Street Journal* has offered a start, however, by giving front-page coverage to Blue Corn, a handsome, genial, fifty-year-old Tewa Indian potter of San Ildefonso Pueblo in New Mexico. Blue Corn has lived simultaneously in two divergent and symbolically opposed worlds. She once served as a domestic to the late atomic physicist J. Robert Oppenheimer, during the Los Alamos experiments on the atomic bomb. An innovator in reviving polychrome pottery in styles known in the Americas before Christ, Blue Corn is being drawn into an increasingly brisk market for pottery which shows the mark of the human hand. Anita Da, the owner of a San Ildefonso studio which markets pottery, commented: "Now mass production is depleting the authenticity of traditional designs—everybody wants to be an Indian."[6] Vine Deloria would agree: "There are many things happening today that can be related to ideas, movements, and events in Indian country—so many that it is staggering to contemplate them. American society is unconsciously going Indian. Moods, attitudes, and values are changing."[7]

Some Personal References

Such observations make a great deal of sense to me, a member of the WASP minority who long ago decided the label didn't fit. Born in Yale, Oklahoma, a village settled by my Deep South maternal grandparents, who migrated to Indian Territory on a covered wagon, I learned as a child in the Roaring Twenties that something was missing: I had no Indian blood, but I might have had if my mother had married Jim Thorpe, her next-door neighbor, who used to court her by jumping flatfooted over the clothesline in the backyard, rehearsing at the same time for his later fame in the Olympics. We learned our fractions in elementary school by asking our classmates, "How much are you?" Everybody knew that this question referred to "Indian blood." Being an eighth Cherokee, a fourth Creek, or half Seminole—any combination—was considered better than having no Indian blood at all. Our two representatives in the Hall of Fame were, after all, Sequoyah, inventor of the Cherokee alphabet, and Will Rogers, the part-Cherokee humorist-cowboy. Through that tribe, we learned that our state had roots far away in the American East—in the mountains and woodlands of North Carolina. Through classmates who were Creek or Sac or Fox, I found links to other "Old World" cultures, Alabama and Iowa.

Tennessee was added to my roster of cultural homelands through my father, who was born in the proverbial log cabin in Palmyra, Tennessee, descendant of settlers who already had tried to eke out an existence alongside Indians before Andrew Jackson sent them both moving along their respective trails of tears looking for a better life. Years later, during the Chinese revolution, I had dinner in Shanghai at the Press Club in Broadway Mansions overlooking the Bund and the Yangtze River. Father O'Connor, Roman Catholic priest-news correspondent, Seymour Topping, another journalist, and I were discussing the various tribes and religious groups, including Jews and Moslems, who made up the Chinese population. Diversity had not prevented the Chinese from feeling superior to all other civilizations.

Feeling proud of his own Irish variant on European civilization, Father O'Connor asked me, "Dillon, my boy, how can you have a name like that and not be of the Church?" I mischievously replied, "What Church, Father?" His question in 1947, when I was well into my twenties, became a trigger for my own "identity search," for it was the first time I had ever heard that my name was Irish. (It is also French and is to be found in India, but I had never thought of my European origins.) Having never heard of St. Patrick, I was like a Jew who had never heard of Israel. I began to resent all that I had been missing and, eventually, started writing an essay, "On Learning To Be Irish: The Case of the Late Bloomer," after having been invited in 1962 by Conor Cruise O'Brien to stay as his guest in the Irish Club of London, where I was accepted for the first time on the basis of an ethnic identity.

Upon returning from the Far East in 1948, I asked my grandfather where we were from. "You know we are from

Palmyra, three generations born under the same roof." "But where before that?" I asked. "North Carolina." "No, originally—what country?" "I'm not quite sure," he replied, "but I think it was Virginia."

That dialogue took place before I learned that the identity game, normally played by tracing kinship and territory through patrilineal descent, ought to include some other rules: finding out something about one's mother's family, and grandparents on both sides. Later, by taking on the relatives of one's wife or husband, the chances for multiple identities start to pyramid. Women's liberation movements seem to have missed a good plank in their platform when leaders ignore the point that a person's legal name, inherited from a father, is but a fraction of the possibilities of following Secretary Ripley's bidding: celebrate yourself, know who you are and where you came from. Thus I am now at work finding out about the Jordans, Hollands, Darwins, Hardcastles, and Gwathmeys, to name a few, who carry me beyond the Irish into other tribesmen, such as the Welsh and English, who make up the British isles. A real test of what one has learned about one's past lies in making an "identity kit" out of fact and fiction, and communicating to one's child some notion of who he or she is. My own son, born in Ghana, albeit of white parents, has a great advantage over his parents in claiming an identity with the continent on which the human race presumably originated. The ease with which he can assimilate the idea of "One Species—Many Cultures," however, does not preclude his enjoying various identities, not all of them ethnic. (His greatest chosen identity at the moment is being the youngest member of the Steamship Historical Society of America.)

The regional element in the American experience represents another important basis for diversity. *The Education of Henry Adams* gives some beautiful commentaries on how the personalities of northerners and southerners varied as the country was about to become 100 years old; the slavery issue seemed not to make the vital difference. I had to learn from direct experience, rather than from Adams, however, that all northerners are not Yankees as I had always been

Wilton S. Dillon

told in the South. This revelation came to me in New Haven when a Jewish classmate of mine from Berkeley, showing me his hometown on my first visit to the American North, pointed out the streets on which the Yankees lived. "But, you're a Yankee, and you are not including your street," I remarked. I was late in discovering the gross limitations of my southern perspective, a tradition which, paradoxically, I now embrace with a greater sense of belonging because of my other belated awareness that having lived in a kinship-oriented, preindustrial society prepared me better for living in a larger universe—Asia, Africa, and rural Europe—than had I known only a northern, urban, social-contract kind of environment. I have come also to realize that southern whites are no less carriers of African culture than those whose skins announce their ancestry. Africa, indeed, is one of my motherlands, and I am richer for having transcended culturally the gene pools of Ireland, Scotland, Wales, and England, and those subsequent regrouping of genes in the mountains of Appalachia and on the farms of Mississippi.

I have volunteered these personal references in the hope of enticing the reader of these essays to make his or her own personal sense out of them, trying everything out for relevance, or size, vis-a-vis personal, social, national, and international experience. Use yourself as data. With the proper rubrics or tools of analysis, your individual or social experience illuminates the abstract theories about culture change and how America has moved from a melting-pot model toward a pluralistic one. "The basis of all culture change is located in changes in the attitudes and behavior of individual members of society," Evon Z. Vogt writes, in calling for a mixture of macrocosmic and microcosmic models of culture change.[8] If not oneself, other contemporary people will do nicely.

In speaking not long ago to the Maryland Chapter of the Daughters of the American Revolution on "An Anthropologist Looks at Ancestors," I passed on Margaret Mead's suggestion that we old Americans ought to share our ancestors and not dwell so much on bloodlines, recalling a British

Tourist Office ad which said, "If you do not have any English ancestors, why not invent some?" I mentioned that Thomas Jefferson is a great hero of mine, an ancestor, if you like, by choice, from among the pantheon of fathers of our Republic. Politically speaking, he is no less my ancestor than those who have the right to be buried in the family cemetery at Monticello. I further encouraged the hospitable ladies of the DAR to join and lead a new cultural revolution by women in the United States who care about children and their understanding of the rights and duties of citizens under our Constitution. I suggested that we old Americans have something to learn from new Americans like the Nader family, who take quite seriously the requirements of our civic ethos—that individuals must try to make the system work by holding government and other institutions accountable.[9] Newness can bring hybrid vigor. Having been around for a long time does not necessarily produce the deep roots or sense of the whole or caring for "all the people." Religious traditions abound with examples of cyclical vigor brought by new converts. Billy Graham once urged the Japanese and Koreans to pray for the people of New York during his rallies in Seoul and Tokyo. Now the Rev. Sun Myung Moon of Korea is trying to bring a fresh look at Christianity back to the American scene as a new form of yin and yang. Similarly, within our own political territory, the DAR traditionalists, guardians of the Constitution, should rejoice over civic inspiration from new Americans like the Naders, fresh from the geographical nursery of the Judaeo-Christian tradition in the Middle East, which once animated our secular political institutions by focusing on the importance of the individual.

The Individual and The Nation

It is the *individual*—whether monocultural or bi- or tricultural—who remains central to an understanding of the drama of change in our national culture as we have moved from a "melting pot" national self-image to a more fragmented or plural one, not yet coded or articulated. Peter Berger, the sociologist of religion, offers in this book a defense of marginality. His essay reminds me of a psychiatry seminar of Dr. Karl Menninger I once attended in Topeka in which he explained his preference for recruiting psychiatrists who possess some degree of ethnic, religious, or social class marginality—"you have to have at least bifocal vision if you are to get outside of your own problems and into those of the patient." Berger writes that alienation is not only the price but the necessary condition for self-discovery in modern society. Take that proposition, apply it to oneself, and then to one's nation and to that subjective and objective set of bonds which tie the individual to his nation-state.[10]

How is one's individual self-esteem protected from the vacillation of one's nation in the world community: the moral scorn of most of the world, for example, against our good intentions in Vietnam? Or, on a more materialistic basis, how does one adapt to the decline of the dollar to the extent even that, in the Jack Lemmon-Billy Wilder film *Avanti*, Italian con men are shown refusing to take their money from an American in dollars, preferring either German marks or Japanese yen? When does pride in the soundness of the dollar define one's respect for oneself?

What happens to the self-respect of individual patriots in any country who, having identified with their government, find the state morally bankrupt or repressive? Both the individual and the nation need to be prepared for role-reversals and sudden changes in status. The recent graceful withdrawal of the British from empire may be a model to emulate, just as one might admire the long-run transition of the Swedes from military conquest to the cultivation of "peace" as an industry. In the American South, family status often was independent of financial solvency. (Perhaps being "poor but proud" is more universal than I think.) Could southern families become a partial domestic model for other Americans as the energy crisis forces us to change our symbols of success and living standards (automobiles and a clipped suburban lawn) and seek new satisfactions from personal definitions of "the quality of life" that might match styles of fifty years ago? Do we need new definitions of success, progress, honor, shame, or consistency in coping with changing times? Need we have—in the face of Kenneth Clark's TV series, *Civilisation*—"shame" for belonging to a nation only 200 years old, "without a history"? Or, if having developed pride in a nation known for its youth, hope, trust, and vigor, how do we adapt to another image which would include (a) the antiquity of our Constitution, compared to others, and (b) the growth of "European-like" suspicion and spying among a population made physically ugly by overeating junk food and reliance on alcohol and drugs to veil us from stress and reality? It is too much to expect that a self-renewal program based on cultural and ethnic pluralism will give us a new national body-image that matches our almost forgotten greatness as a people. That would be putting too heavy a burden on that provisional goal I playfully suggested earlier: "Be American, Be Different!" However, because I believe a personalized understanding of the essays here provokes us to think hard about ourselves, and thus oppose the changes that drive some of us to abandon democracy for a quest of security and certainty, I feel obliged to look at the pros and cons of celebrating diversity and differences under the banner of *e pluribus unum*.

Limitations on the Idea of Diversity

In the wake of the attempted assassination of Governor George Wallace, the columnist Joseph Kraft asked: "Why? Oh, why?" and said the answer lies right before our very eyes. "During the past decade, the country has been on a binge of unbridled self-expression. Every group, and virtually every person, has been staking claims that crack established supremacies and traditional restraints. Given an open season on authority, young people are the obvious starting point. The new life style they have invented includes a new kind of music, a new form of dress and hair style, different attitudes toward success. But in fact the young assert their values in ways that are not benign. They undermine the chief restraint on Western society—the restraint of conformity, which is another way of saying respect for other people and their values."[11]

Kraft, the discerning social analyst who also brought us essays on "middle America" and the "forgotten majority," added this caveat about the celebration of ethnicity: "Italians, Jews, Chicanos, Greeks, Slavs, and the various Orientals are now advertising their origins with a vengeance. Ethnic narcissism sets group against group in an invidious competition—a competition particularly harmful to blacks. It also saps another important restraint on behavior, another concept fostering respect for the majority. That is the concept now derided as "assimilation."

He then links violence to the loosening of older bonds. "Most Americans can find a place in this tumult of Meism. It is the genius of this pluralist country to offer a pluralism of ego satisfaction. But there are misfits cast adrift from normal participation in group life—persons unable to keep up with the dizzy pace of American life. These shattered beings find the means of self-expression in lonely acts of desperate destruction."

Demands for the right to welfare and quotas in jobs and university admissions look like special privilege, with everybody pushing claims and refusing to see the whole. Prof. Benjamin Ringer of the City University of New York is pushing, in this context, his own claim as a sociologist: the need for worldwide research—under UNESCO auspices—on how ethnicity fits into occupational structure. If, as Peter Berger says in this book, society assigns identities, is not the getting of a job the central process to examine in finding out how ethnic, racial, or religious factors may shape staying alive? The blood of Belfast flows across a kind of artificial (nongenetic)—or socially inherited—"ethnicity." Catholic or Protestant "identity" figures in the competition for scarce jobs or votes. Nobody could claim Ireland as a pretty model for celebrating diversity. The killing also destroys chances to use the healing qualities of poetry, proverbs, and folklore to find some common denominators. Sharing the same God and Celticism counts for little. In some societies, ethnicity is more than a leisure-time occupation, as witness, also, the place of Jews in the Soviet Union.

If one emerges confused after reading some of the essays in this book, never mind. The contributions were not intended to fit together as a syllogism. Neither do the elements of the times in which the essays were produced. Social units defy coordination by central policies. Theodore Lowi's *The Politics of Disorder* [12] suggests that politics and public life were once almost synonymous, but increasingly government seems a private affair. Could the recent moves toward the privacy of a cultural home base (where one can enjoy soul food appropriate to various traditions) represent an emulating of those who claim secrecy or confidentiality as essential to statecraft? Has government become, like Greeley's intellectuals, an ethnic group in itself?

Since Alexis de Tocqueville, Americans have been accustomed to borrowing pieces of our self-image from French and other foreign observers of the American scene. From his 1832 *Democracy in America,* we learned of our uniqueness in developing voluntary associations to serve as buffers between the individual and the state. Now, in Alain Touraine's dialectical essay, we find some other propositions to contemplate: he writes of alternations between the integration and the openness of society, the centrifugal pulling together, and opening up, as distinct from falling apart. Opposition to the status quo is extraparliamentary; it is both utopian and violent. Like the maxims of La Rochefoucauld, Touraine's generalizations are so broad that it takes some effort for the reader to see where American individuals or the society fit. Consider our new pluralism, for example, both pro and con, and ponder his statement that "social identity can be born only from participation in the conflicts which form around the control of the processes of social change." He does not refer to the environmental movement, though it might well illustrate the point that some Americans, as though participating in a new religion, show their reverence for life by volunteering to try to reduce air and water pollution, only

to find an uncoordinated conspiracy against their efforts, for the environmentalists, as ecomaniacs, might disturb the economy, our warmaking powers, or our Gross National Product. What President Eisenhower called "the military-industrial complex" becomes identified with saving not its identity, but its present scale of existence, and so sabotages its new "enemies." The conflict thus arises between those who take short-term perspectives and those who feel virtuous for "regressing" in order to save the fragile web of life from possible extinction.

What has this nearly cosmic or biospheric drama to do with homogeneity—or the unmeltable ethnics? Public issues are perceived differently, of course, according to one's attitudes about the whole and the parts. Who can keep his psychic and social energies applied to redressing grievances by the afros and dashikis and declarations of "black is beautiful," while challenging other consumers, the government, and industry to protect the whole environment for the good of Italians, Jews, Chicanos, Greeks, Slavs, and the various Orientals? In the larger world, environmentalists in industrial nations responsibly try to warn the developing countries about the dangers of pollution and population only to discover that new nations may demand the right to pollute —"just a little"—as a symbol of coming of age by standards of the old order. And the newer countries also resent the "neocolonial" or "racial" implications of being advised to produce fewer black or brown children, or to concentrate on exporting raw materials for processing in the societies which already "have it made," industrially speaking. I am only suggesting that low-income ethnics may resist the global morality of persons affluent enough to care about the whole planet and angry enough to spend time "helping others" in addition to those contained within their own ethnic territories. There seems to be little difference in this phenomenon whether analyzed domestically or internationally. (More light will be shed on this with the publication of David L. Sills's *The Environmental Movement and Its Critics* by the Russell Sage Foundation.)

A practical question raised by Touraine is how to enlist greater participation by all cultural groups in "the conflicts which form around the control of the processes of social change." We can no longer be content to say "let the best ethnics win." Indeed, there are stopping points. Pluralism can be dysfunctional, even suicidal and conducive to violence. "Balkanism" figured in "starting" World War I. Basque separatism provided motives for the killing of Admiral Carlos Arias Navarro, the Spanish prime minister, in December 1973.

Still, depending on the level of discourse, the issues need not be debated as either/or. Self-knowledge is a precious commodity, just as self-deception is dangerous. An Alex Haley, who goes back to find his roots in Africa, may very well be the potential environmental leader. Having seen some of the joys of human relationships in his ancestral Africa, he may use his newly reinforced identity as a platform for enlisting multicolored support for protecting ourselves from the evils of unguided modern technocracy.

The turning away from the melting pot has not happened suddenly, and it has seemed to have no central leadership. But a drama has taken place, nevertheless, in the American cultural scene, and we move toward the Bicentennial celebration protesting not a foreign ruler, but proclaiming independence from homogenization and celebrating a continuing revolution: the rights, pleasures, and duties of enjoying several identities while protecting the same aim for others. In this drama, there is no dénouement. The last act has yet to be written.

Human beings, however, like to speculate about the future, worry about it, or hope that something better is about to happen. Thus the temptation to guess about what is going to emerge in "the last act." A look at other human experience, historical and contemporary, may be useful.

Other Models To Ponder

As the United States faces its third century as an organized society, what can we learn about other nations' experience in adapting to diversity within a given political territory, and to whole nations' adaptations to a multicultural world? What are the alternatives to a conformist, mass society governed by technocratic policemen who see cultural difference as a source of subversion or a sign that "if you are not with us you are against us?" Are there any models, historical or hypothetical, which Americans might ponder? The city-states of Singapore, Hong Kong, and, let me add, Manhattan and San Francisco, are fascinating instances of what Henry Adams called multiversity. But do we not need bigger models than city-states? Think first of bilingual Montreal, less for its intrinsic interest as a city than its importance as a concentrated urban manifestation of a larger whole, where the actors in Canada's drama of pluralism can be seen on a single stage. Toronto similarly boasts a mosaic of cultural enclaves. A visit to Montreal for an anthropology meeting gave me the unexpected sensation of being in Johannesburg. The similarity was not the strains between the white settlers and the "aborigines" (North American Indians in one case and blacks in the other); no apartheid for Indians prevails in Canada. The crude similarity, rather, seemed to lie in the stresses between two groups of European transplants—the English and French in Canada, and the English and Afrikaan descendants of the Dutch in South Africa. The gleaming chrome and aluminum and glass of modern skyscrapers in both cities are deceptive stages for the drama of conflict just below—and occasionally above—the surface in Montreal and Johannesburg. Theory and practice are not yet joined when bombs and kidnapping are used to dramatize grievances.

Yet, from the U.S. perspective, Canada deserves special attention despite her preoccupations with two major groups—while we cope with much greater multiplicity below the border. Canada's intellectuals and political leaders have stressed the positive values of diversity to avoid further fragmentation or separatism. The actual situation reveals that countries cannot live by rhetoric alone; leaders there or here cannot verbally *will* harmony by proclaiming the joys of heterogeneity while the allocation of political and economic power along language or ethnic lines remains a bone of contention. Political leadership which invites cultural gifts from all can go a long way toward creating a climate of mutual respect, a civility which encourages both the protection and sharing of distinct lifestyles. Canada's ongoing experiment with pluralism—though we have no sharp equivalent of *Quebec libre!*—may be one of our most instructive models.

We have, indeed, some Indian tribes in common. One, caught between the two countries, the rice-gathering Chippewa or Ojibway, have about 35,000 members in the northern United States and another 30,000 in southern Canada.

The Turtle Mountain Chippewa band near the northern border of North Dakota and Manitoba are North American reminders of the identity processes of the Ewe who straddle Togo and Ghana, the Yoruba who spill over from Nigeria into Dahomey, or the Wooloofs who interconnect several nations on the Guinea coast. At a 1973 meeting in the U.S. Department of State, discussing the role of American Indians in foreign affairs, a vivacious young Chippewa woman, Twila Martin, declared: "We are first Turtle Mountain, then Chippewa, then Indian, and then citizens of the United States." Her Chippewa band had petitioned, in 1959, the Russian ambassador to the United Nations to intervene on behalf of Turtle Mountain to settle some land dispute with the U.S. Government, a task politely declined by the diplomat, who said, "We don't want to interfere in the internal affairs of other nations."

The Chippewa case is interesting in the context of this book—particularly the essays by Vine Deloria and Murray Wax—because it should inspire non-Indians to "try on for size" the world view of a proud cultural minority resisting both "Americanization" and "Canadianization." The Chippewa would project similar tenacity, no doubt, if they found themselves, say, on the border between China and the Soviet Union, or in an Andorra or Liechtenstein situation in Europe. The drives toward autonomy appear to be as universal as the desires of small groups also to participate in and belong to a larger segment of the human family. One need only to recall that Aldous Huxley's *Brave New World* celebrated the durability of the Zuni.

And what about China and Russia as models? The pre-revolutionary and postrevolutionary experiments with pluralism in both the U.S.S.R. and China represent a vast uncharted area of comparative ethnic and political study. The cultural policies of regimes before and after their revolutions would make useful studies in themselves. Then, scholars and politicians might assess also the relevance of both societies for an improved understanding of how diversity, identity, protest, and change operate on the American scene.

In the Taipei museum of the Academia Sinica, in 1972, I saw exhibits of mainland China's prerevolutionary ethnic arrangements, a far greater diversity of lifestyles than I had thought possible in a civilization which had been so long at work consolidating and codifying its cultural components. The Nationalist Chinese were recording things as they were, and revealing striking similarities with their mainland brothers in their perceptions of China's historic patterns of cultural hegemony. Now that American scholars, diplomats, and traders are interacting with Peking leaders and their provincial counterparts, I trust that we shall discover soon the changes in theory and practice of pluralism before and after the revolution. How can cultural differences survive the demands for political unity? Are cultural, linguistic or religious minorities, such as the Uighurs, tolerated so that they can be more manageable? Perhaps novels, plays, ballet, acrobatics, or poems will reveal as much as party tracts or Mao's red book the official position on *e pluribus unum*, Chinese-style. How does a Tibetan, a Chinese Moslem or Jew, a member of the "Five Peoples" in the official flag, or a herding and grazing Kusath, rank his or her citizenship in relation to other identities? Is a member of a Chinese cultural minority comparable to a Turtle Mountain Chippewa vis-a-vis other Chippewas, other Indians, and then citizens of the United States? Are Leninist and Maoist views reconciled on "the nationality question" and "self-determination"? What has ideology to do with defining an identity or shaping a sense of belonging? (An exceptionally perceptive young American Sinologist in our Foreign Service told me that "We Americans often act as though we don't have an ideology; only foreigners are believed to have ideologies, but we *do* have one, and our policies and behavior often reflect it.")

Chinese or Russian interpretations of Karl Marx's attitude toward self-determination could have plenty of latitude. His idea of "nationality" was ambiguously used to refer to "society," "state," or "country." Federalism and states' rights, as worked out on the North American continent, reflected Hobbes and Locke, and antedated Marx. Moreover, they were free from linguistic and ethnic considerations comparable, say, to the ethnic groups surrounded by the Great Russians. Perhaps only the vast nation of the Navajos—the size of the Low Countries of Europe—has some correspondence with an autonomous Soviet Republic. Afro-American "nationhood" has no comparable land base. Had Vermont become Israel, or if Mississippi had been transformed into a black republic, the United States might provide neater parallels to the Soviet experiment. Even without such parallels, historical and contemporary Russian approaches to pluralism help to illuminate what is going on here beyond the melting pot.

For ethnic balance and obeisance to women's lib, consider the proverbial search in Washington for "a black Alaskan nun who speaks Spanish" to put on federal committees. Back in Stalin's early days, he headed the Narkomnats, or People's Commissariat for the Affairs of the Nationalities, which served as watchdog, organizer, sovietizer, and protector of the nationalities until the Soviet Union was formed in 1923. In Moscow today the Ministry of Culture takes over some of these functions and carries the symbols of the multinational state into public entertainment. I recently saw the Moscow circus. Resembling the format of a Broadway musical, the circus entertained and educated at the same time. Its audience could not escape the highly explicit diversity principle manifest on the bars and trapezes as Georgians and Mongols were heralded with trumpets to do their cultural or ethnic thing. The circus's universe extended to East European allies so that the audience was treated as well to the antics of Bulgarians and Hungarians. In Warsaw I watched the Poles celebrate the anniversary of the Russian revolution by the performance of a multiethnic folk dance and song group, following the political speeches. The National Unity Front recognizes such diversity as appropriately Polish. Though the ethnographic museum has no sampling of Tartar culture surviving in Poland, one sociologist told me of a pocket of them "alive and well" with their double identities.

Ideologies and political differences aside, the scale and support of Russian scholarly efforts to study ethnic phenomena today deserve emulation in the United States. Comparable to studies of their own society, Soviet ethnographers have produced a cultural and linguistic map of the United States showing the distribution of various Indian language groups irrespective of political boundaries. And in studies led by Dr. Yu. V. Bromley, the Institute of Ethnography of the Soviet Academy of Sciences has excelled in providing a theoretical framework for analysis and interpretation of the persistence of ethnicity and separate identity in modern times, including the fascinating cases of Soviet Georgia and Armenia. While more "structural" than "psychological," the Russians make their contributions to understanding identity formation by pointing out that racial distinction and physical appearance make little, if any, difference in explaining ethnic "bonding." Religion, language, folk art, folklore, customs, rites, and norms of behavior are far more pivotal in forming a group consciousness. Bromley recently has studied the impact of the mass media and army service on the twin processes of becoming a more integrated society and the increase of ethnic awareness by being exposed to greater diversity.

Were the United States able to emulate the Soviet research apparatus, if not the substance or findings of ethnographic investigation, the Smithsonian might well be the institution most appropriate to serve as a national and international center of work on ethnic communities. (Americans already owe the Russians a debt for stimulating our space research and primate studies.) We would then be in a much better position to become a "trading partner" of the Soviets, the French, the Chinese, the Mexicans, the Malaysians, the Indonesians, and any others whose work on human cultural adaptation would help us understand better the idea of "one species, many cultures." The trade would be reciprocal. Politicians and scholars in Africa, Asia, Latin America, and Europe all have a stake, too, in keeping up with the U.S. experience. In one way or another we are all "a nation of nations," still trying to create political communities compatible with mixed individual and cultural identities. The growth of the global society and the individual's love for the whole planet rest on the self-respect of small groups writ large. It is the modern secular equivalent of the Golden Rule: love yourself before you can love others.

In this primitive effort of mine to make a random sweep

of the earth to examine, ever so prematurely, other experience to enlighten us about our own move toward a diversified model to replace the melting pot, I have touched upon Canada, China, and the U.S.S.R. when I could also have used profitably Indonesia, South Africa, India, Malaysia, Israel, Spain, and Yugoslavia for heuristic purposes. In all of these we would have found evidence of implicit trade-off relationships between differing groups within the same political community (e.g., the political power of the Malays balanced against the economic power of the Chinese in Malaysia). So, for any future, large-scale comparative exercise we would need to look at the dynamics of interdependence and reciprocity in a given society. A most useful point of departure for future scholarly efforts, aimed at both new knowledge and applied political science, is Clifford Geertz's "The Integrative Revolution" in his collection of essays, *The Interpretation of Cultures*.[13] But none of the existing literature, including the essays in that book, has yet dealt, in the same context, with two ends of a spectrum—from heterogeneity to homogeneity—at opposite sides of the Pacific basin: New Guinea and Japan. Sam Stanley, a Smithsonian anthropologist, when asked what experiments with cultural pluralism he thought would be instructive for the United States, immediately responded: "New Guinea. It's out of this world; the building of an independent state out of that diversity would blow your mind!"[14]

The Japanese experience, of course, has been "built" for a long time, and Japan would hardly qualify to appear in a list of plural societies. Glazer and Moynihan in *Beyond the Melting Pot* observed that Japan finds it impossible to incorporate into the body of its society anyone who does not look Japanese, including Koreans. (They also wondered how Russians would respond to a testing of racial attitudes if they should experience a large number of blacks living among them, resisting Russification.)

Nevertheless, the Japanese experience is most interesting and significant for our purposes. Paradoxes are instructive. Though not comparable to the continental scale of the United States, China, or the Soviet Union, Japan is a technological and scientific giant among nations (even with the energy crisis), and contains great homogeneity, stratification, and various forms of protest deserving our attention if we are to learn more about ourselves by examining others.

Japan, the homeland of the chairman of our symposium, Michio Nagai, has been described by Kenneth Boulding, the economist, as the first country of the twenty-first century. (He wrote before the energy crisis dramatized the hazards of being such a pioneer.) Charles F. Gallagher, the cultural historian, refers to Japan as the first totally urban society, where distinctions between town and country have been blurred by rapid transit, mass communications, and rural electrification.[15] As described by Chie Nakane, moreover, Japan may be the most homogeneous of any nation-state.[16] France, with all her state centralism growing out of Louis XIV and Napoleon, cannot compare with the sinews and bonds which reinforce the webs of Japanese society, with its preindustrial paternalism alive and well inside labor unions and modern factories. Mishima's suicide, as interpreted by Donald Keene, represents the ultimate protest of an individual against his nation's allegedly deserting its traditional values in favor of a culture governed by Gross National Product and the internalization of the way of the machine instead of the way of Tao, the tea ceremony, and Bushido. Mishima, with whom I once dined in a Chinese restaurant in Harlem, exemplified the links between identity, diversity, and protest.

In his violent symbolism, Mishima represented also the paradox of the cosmopolite who had tasted the whole world in its infinite variety before coming back to embrace the most traditional of classic values. Thomas Jefferson (happily not a suicide) was an earlier example of a man with a vision of a larger universe who brought much of it home with him to cultivate local traditions.[17] Marginal as Mishima may have been, even to a culture with a personal honor suicide tradition, he shared with the most insular and parochial of his fellow citizens a curiosity about the outside world and a dependence on it for stimulation. Japan, deprived of internal ethnic diversity, seems to compensate with an insatiable hunger, satisfied by electronic means and affluence, for the diversity which persists elsewhere in societies not yet so smoothed over by shared patterns of culture. Rich have-nots can import.

Though racially and culturally homogeneous, the Japanese have transcended the East-West dichotomy and are increasingly lumped with the "Western World," in view of science, technology, commerce, and, yes, musical tastes. Like the waves of an ocean reaching a beach, Japan is spreading her distinctive style and, in the process, bringing back with each wave the material culture of other lands. The fine tuning mechanisms, the cultural antennae and radar which prompted the Japanese to accept Hollywood standards of nose length and bosom size for women and to know who or what is Number One on the Hit Parade or the Book of the Month Club, suggest to me that such outward looking need not be done at the expense of one's inner self. Japanese communications with other cultures—communications, not imitation—serve them well in picking up cues even about the ethnicity boom abroad. I have examined Japanese-made "Indian" war bonnets and moccasins in the Tulsa airport and souvenir counters at Gay Head, Martha's Vineyard.

If this glimpse of Japanese and other societies allows one to examine the American experience as one point on a scale running from homogeneity to heterogeneity, the moral of the comparative exercise might be that (a) no nation has yet found "the answer," (b) we can be thankful for our own diversity, and (c) we should avoid knocking technology as a bad, homogenizing influence in itself, for it depends on what we do with technology—whether or not it is used to enrich our lives. The mass media are neutral; they can flatten out and homogenize priceless differences, or they can facilitate their sharing. (South African officials have been slow to install television, allegedly fearing it would "stir up" the blacks, or that programming would be too difficult for so diverse a population. The United States experience, on the other hand, shows that "ethnic tuning" on a TV or radio set allows a viewer or listener to pick and choose what suits his ethnicity, "class," or mood—a kind of audio-visual cafe-

> **E NEW YORK TIMES, MONDAY, SEPTEMBER 17**
>
> ## Ethnic Fetes Tete-a-Tete
>
> **By JUDITH CUMMINGS**
>
> It was standing room only yesterday where Little Italy and Chinatown stand back to back as the procession of San Gennaro, on the fourth day of the Italian feast, wound from Mulberry to Mott Street and smack into a Chinese culture festival.
>
> The parade, honoring the patron saint of Naples, acquired an unplanned cross-cultural flavor when it drew alongside Columbus Park. Without warning, the sounds of Chinese music boomed above the red, white and green floats, momentarily drowning out the Italian brass bands and causing not a few surprised looks among bystanders.
>
> "It just works out that way sometimes," said Howard Lew of the Chinatown Planning Council, who had just finished an opening speech for the third and final Chinese festival of the season in the park.
>
> "We didn't know it was going to fall on the same day as the San Gennaro festival," he said. "But they've been around the same time for the last three years."
>
> The coincidence provided a politician's field day, and more than one candidate for city office who had been marching behind the floats took time out from the parade to say a few words to the predominantly Chinese audience in the park.
>
> Robert Jordan, who lives in Port Washington, L. I., and said he did not often get to come into the city, was one of the visitors who was caught off guard.
>
> "I told the cab driver to take me to Chinatown," he said. "When I saw the Italians going through the middle of it, I said 'Hey, this is pretty rotten,' until I found out they were having a festival too."
>
> "This is like a bonus," he said, threading his way past the sausage and pepper stands on Mulberry Street.
>
> On the other hand, Grace Hunnicutt and Christina Ricardo has known that the two festivals were to coincide and had made the trip because of this.
>
> "It's like a cultural exchange," said Miss Hunnicutt, a legal assistant for the Hartford insurance group. "That's what the problems are all about—people don't take the time to appreciate each other's culture."

teria. Washingtonians, black and white, for example, can spin the dial to AM or FM and hear Bach, Beethoven, soul, rock, blue grass, or country. The mass media can either widen one's tastes and horizons, or yield to one's exclusive ethnic preferences.)

We in the United States, then, may be mentally healthier and have the prospects of greater unity on our 200th birthday if individuals and groups are encouraged to celebrate their distinctiveness, and to rediscover lost, submerged, or abandoned cultural identities which reveal our genetic and cultural links to other times and other places. We must remem-

Wilton S. Dillon

ber that individuals need not be confined to just one cultural pigeonhole. *E pluribus unum* can drop the question mark when we develop a poetic, rather than a literal, understanding of the idea of "one," and know that one is plural. The growth of a global society transcending race and ideology has not yet made nation-states totally obsolete as experiments in community. Until then, individual nations need to learn from each other the evolving rules of good cultural housekeeping: the wise management of pluralism so that a society's "need" for violent protest, civil war, or assassination is reduced. (Nations also can learn from institutions within their borders, as witness the case of the American soldier threatened with court martial for wearing the beard and turban of his newly embraced Sikh religion. Can or should whole societies be so regimented?)

With his civilizational notions of "the raw and the cooked," Claude Lévi-Strauss demonstrates that culinary metaphors may be useful in thinking about a new, shorthand description of pleasing social arrangements which could supplement or replace the melting-pot crucible. We have seen that *e pluribus unum* itself may have had a culinary—or floral—origin (page 39). Perhaps in 1976 we should give prizes in a contest devoted to suggestions for "beyond the melting pot." Can each nation, and then a world society, design openended cultural menus which recognize diversity *and* solidarity: a fondue, an omelette, fruit or chef's salad mixed to one's own taste, meat and potatoes, chop suey or shishkebab?[18] Some have invested so much in the way of adaptation or assimilation that a return to the status of "hyphenated American" may seem like regression. Such "homogenized" Americans should be encouraged to follow the Japanese example and tolerate diversity of outsiders, for their own freedom to remain monocultural may depend on that tolerance. And if the imagery of food does not appeal, music might. The African teacher and philosopher, J. E. K. Aggrey, reminded listeners on both continents that the black and white keys of a piano are both necessary to produce melody. Without such metaphors of music, food, religion, or politics to hold us together while we seek dignity and selfhood in our own ways, or to continue to think happily of this earth as home. Martin Luther King, Jr.'s tombstone bears these words: "Free at last, free at last, thank God Almighty, I'm free at last." Did his sense of freedom spring, in part, from his metaphysical escape from the confines of America's cultural categories? Hopefully, his survivors will not have to die to feel free.

NOTES

[1] NATHAN GLAZER and DANIEL PATRICK MOYNIHAN, *Beyond the Melting Pot* (Cambridge, Mass.: MIT Press, 1963).

[2] ANDREW GREELEY, "Intellectuals as an Ethnic Group," *New York Times Magazine*, July 12, 1970. Greeley is the founding editor of *Ethnicity: An Interdisciplinary Journal of the Study of Ethnic Relations* (Academic Press, New York).

[3] MICHAEL NOVAK, *The Rise of the Unmeltable Ethnics* (New York: Macmillan, 1972).

[4] In a talk before the National Press Club, Washington, D.C., October 12, 1973.

[5] See special issue, "Ethnicity and Social Work," *Social Work*, volume 17, number 3, May 1972. This issue contains such articles as Nathan Glazer's "Interethnic Conflict," and Joseph Vigilante's "Ethnic Affirmation, or Kiss Me, I'm Italian." Also see William D. Davidson, M.D., "Self-Image and Social Reform," unpublished paper presented before the American Psychiatric Association, 1968.

[6] MIKE THARP, "The Craftsmen," *Wall Street Journal*, September 14, 1973.

[7] VINE DELORIA, JR., *We Talk, You Listen* (New York: Macmillan, 1970). Another example of the reversal of talking and listening roles is found in the *New York Times*, April 30, 1972, report of a new Indian-owned radio station in Ramah, N.M., broadcasting to 1,500 Navajos.

[8] EVON Z. VOGT, "Culture Change," in *The International Encyclopedia of Social Sciences*, edited by David L. Sills (New York: Crowell-Collier Macmillan, 1963).

[9] See Thomas Whiteside's profile of Ralph Nader in *The New Yorker*, October 8 and 15, 1973, for some evidence of the intensity with which Nader responds to his perceptions of the duties of American citizenship, a conception he learned both at the table of his Lebanese-born parents in Connecticut and in the Harvard law classes of Roscoe Pound.

[10] Without solving all of the methodological problems in studying "national character," I once tried to trace a set of such bonds in a portrait of a Gaullist in my study of reciprocity, *Gifts and Nations* (The Hague: Mouton, 1968).

[11] See "The Reason Why," *Washington Post*, May 21, 1972.

[12] THEODORE LOWI, *The Politics of Disorder* (New York: Basic Books, 1971).

[13] CLIFFORD GEERTZ, "The Integrative Revolution" in *The Interpretation of Cultures* (New York: Basic Books, 1973).

[14] Such problems are suggested by Perry Josef Pataki in *Community, Time, and a New Guinea Landscape* (Seattle: University of Washington Press, 1972) and *Politics in New Guinea*, edited by Ronald M. Berndt and Peter Lawrence (Seattle: University of Washington Press, 1973).

[15] CHARLES F. GALLAGHER, "Urbanization, Integration, and National Society, The Japanese Experience," a paper read at an American Universities' Field Staff Conference, "Urbanization: Freedom and Diversity in the Modern City (University of Albama, December 12-13, 1968).

[16] CHIE NAKANE, *Japanese Society* (University of California Press, 1972). See also Hitoshi Watanabe, *The Ainu Ecosystem: Environment and Group Structure* (Seattle: University of Washington Press, 1973).

[17] WILTON S. DILLON, "Thomas Jefferson on Home-Grown Intellect," *Phelps-Stokes Fund Occasional Paper*, 1962.

[18] Serious students of such metaphors should consult, of course, the types of societies and civilizations evoked by Claude Lévi-Strauss's concepts of "the raw and the cooked," or the social equivalent of "boiled," "smoked," or "roasted." See his "Le Triangle Culinaire," in *l'ARC*, number 26, pages 19-29, (Revue Trimestrielle, Aix en Provence).

Scenarios

Actors, Roles, and Stages

Conor Cruise O'Brien
Member of Parliament, Ireland

Statesman and author. Specialist in modern literature and history. Member of Irish Parliament since 1969 and Minister of Posts and Telegraph since 1973. Various posts, United Nations Organization, 1956–61, and vice-chancellor, University of Ghana, 1962–65. Former Albert Schweitzer Professor of Humanities, New York University, 1965–69. Author of Parnell and His Party *(1962),* To Katanga and Back *(1963),* Writers and Politics *(1965),* United Nations: Sacred Drama *(1967),* Murderous Angels *(1968),* Power and Consciousness *(1969).*

*Did that play of mine send out
Certain men the English shot?*

W. B. Yeats, when near to death, asked himself that question about his play *Cathleen Ni Houlihan*, first performed in 1902 with Maud Gonne in the title role. The "certain men" were those who took part in the Rising of Easter, 1916, in Dublin.

In that Rising, a woman took a leading part; the metaphor is here not a dead one as you will see. The woman was Constance Countess Markiewicz, born Constance Gore-Booth. Yeats had known her since his early youth in County Sligo. The poem that opens:

*The shades of evening Lissadell
Two girls in silk kimonos both
Beautiful, one a gazelle*

is about her and her sister, Eva. Constance was sentenced to death for her part in the Rising; later reprieved. From her prison cell, in Aylesbury, she wrote that autumn to her sister, Eva, about Yeats's *Cathleen Ni Houlihan*. "You remember," she wrote, " 'They shall be remembered for ever.' " That was a line in the play; it was the prophecy made by Cathleen, the allegorical figure who personifies Ireland, about those who took part in the Rising of 1798. Constance goes on in her letter to her sister: "What we stood for, and even poor me will not be forgotten and"—she quotes from the play again —" 'the people shall hear them for ever.' That play of W. B.'s was a sort of gospel to me"—and she quotes from it again and finally—" 'If any man would help me, he must give me himself, give me all.' "

The author of the play thought about the prisoner; he had long been estranged from her, and detached from the mood in which he had written that play. He heard a story of how a seagull had come to the window of her cell, and how she had fed it.

Did she in touching that lame wing
Recall the years before her mind
Became a bitter, an abstract thing?

We know that she recalled, out of those years, his play, with its summons to totality of sacrifice, and its promise of remembrance. The mind which had become in his eyes "a bitter, an abstract thing" was the mind in which his words of fourteen years before were imprinted as "a sort of gospel."

We know that she was not the only one of her generation to be affected in this way by *Cathleen Ni Houlihan*. Another revolutionary who saw it, P. S. O'Hegarty, called it "a sort of sacrament." And a spectator who disapproved came away from a performance asking himself "if such plays should be produced unless one was prepared for people to go out to shoot and be shot."

To the question of the dying poet—

Did that play of mine send out
Certain men the English shot?

—it seems that the probable answer is: "Yes, it did."

What is the difference between the play *Cathleen Ni Houlihan* and the Rising of 1916?

Can we say flatly that one is "fiction" and the other "real life"; if participants in the "real life" action have taken in the "fiction" as a "gospel" or "sacrament," is that a transforming spiritual agency?

If a participant, *remembering* the play and remembering its conditional promise to be remembered forever, enacts the

Conor Cruise O'Brien

sacrifice demanded in the play, is she not in fact continuing the action of the play, in terms which are different, not in the sense that real life is different from fiction, but in the sense that the new action crosses a social and conventional threshold into illegality and violence?

It was legal for Maud Gonne to step onto a raised platform in a concert hall in Dublin, pretend that she was Ireland, and utter words which were a summons to violent action.

It was illegal for Constance Markiewicz to step into a park in Dublin, announce in effect that she was one of Ireland's children summoned by her to her flag—which was the language of the Easter Week Proclamation—and use firearms to defend the park and the flag against the same forces which Maud Gonne had symbolically defied.

Constance, therefore, played *Cathleen Ni Houlihan*, using techniques which were far more obviously drastic than those used by Maud Gonne. These techniques entailed the gravest direct personal risk for herself and others. But Maud Gonne's performance entailed the same risks, only indirectly, insofar as she transmitted the "gospel" to the future martyrs (of whom she would have liked to be one herself).

So the essential differences between the two transactions remain in the legal and technical orders. In part of "reality" they are on the same footing, both real and both symbolic. For, of course, the Dublin Rising of 1916, which presented no hope at all of military victory, was a symbolic sacrifice, deliberately designed so that its actors, like the men of 1798 evoked in *Cathleen Ni Houlihan*, should be "remembered forever," and so that their memory, like the play itself, should be an incitement to further sacrifices.

I have talked, in Northern Ireland, with certain men whose sole ambition it is to play such a role—in Constance's terms, not Maud's—and be among those who are remembered in this way. For these men, though they use political language, what is important is not a political objective: the idea that their professed political objective is almost certainly unattainable not only does not deter them, it does not even interest them much. What interests them is the identification with the role of the patriot rebel, and a need to be remembered in this role: the posthumous audience is important.

This is what gives meaning to their lives: it is their honor. And nothing can make their performance in the role memorable except the dimension of risk, up to the supreme sacrifice. Without this element it would all, as they think, be "play-acting": with this element, it becomes "real."

Yet it is real, and it is also play-acting, at one and the same time: resembling in that also an action on the stage.

It would be a mistake, I think, to regard these phenomena as purely or especially "Irish," although it is true that subject peoples have somewhat more incentive than others to explore the possibilities of symbolic action, certain other forms of action being usually closed to them. In Sicily the puppet theater helped to establish the conventions of the Mafia. But peoples can be "subject" in other ways than by nationality, and the history of dispossessed elements among dominant peoples shows elements comparable to the relation of *Cathleen Ni Houlihan* to Constance Markiewicz. The first night of Beaumarchais's *Marriage of Figaro* was an episode—a real episode—in the French Revolution, and the dialogue of Bazarov in Turgenev's *Fathers and Sons* set the tone for more than one generation of Russian protorevolutionaries. The plebeians are always, in some sense, rehearsing the uprising.

But it is not only in the revolutionary tradition that the stage and reality mingle. The play—including plays of so-called "real life"—can incite to violence, but it can also be a substitute for violence, a form of redirected activity. The United Nations is a theater for and of activity of this kind. Again and again, the stages of the Security Council and General Assembly have been used for symbolic representations, once or twice to legitimize the use of force, but much more often to sanctify its avoidance. In all cases the basic political decisions were taken elsewhere, but the United Nations was the scene for a spectacle making these decisions acceptable and, in a secular way, holy.

In the case of Korea, the Security Council chamber was the scene for a rite of sanctification, accompanying the American decision to intervene. This rite was not, of course, strictly speaking, a Security Council decision at all: Article

27 of the Charter requires "the concurring votes of the permanent members" and one of these, the Soviet Union, was absent and did not concur. Legalities were, however, in practice unimportant; the spectacle mounted in the Security Council chamber was adequate for its purpose: symbolic drama of legitimation, propitiation, and reassurance, comparable to the pouring of a libation. This ceremony surely helped the American public to feel that Korea was a just war: a part of the uneasiness felt about Vietnam was that it had gone unblessed, since it was not possible, in contemporary conditions at the United Nations, to stage a ceremony analagous to that performed just after—not before—President Truman's decision to intervene in Korea.

In the case of Hungary, the United States used the United Nations stage for a ritual drama with the reverse significance to the Korean one. The Eisenhower government, pledged though it was to "roll back the Iron Curtain," decided in fact not to come to the aid of the Hungarian rebels, but instead to present a cleansing and compensatory ritual drama at the United Nations. This depended, for much of its effect, on imprecise public recollections of the Korean performance. *Then* the Security Council had *acted*, in a real or literal sense, so the public thought. *Now,* in the case of Hungary, the Security Council "was powerless to act," because of the Soviet veto, and because the Soviet representative was present this time. The spectacle of "the Security Council powerless to act," because paralyzed by the Russian veto, was in fact the *cleansing* scene in the ritual drama. It was cleansing by means of a scapegoat stricken with paralysis. Eisenhower and Dulles were freed of the guilt of allowing the Hungarians to be crushed: the United Nations bore that guilt. The *compensatory* scenes of the same drama were staged in the General Assembly: speeches and resolutions condemning the Soviet Union for its action in Hungary. The representatives of the United States could be seen and heard by huge television audiences, playing active and virtuous parts in these scenes. There can be no doubt that the availability of these dramatic resources made it easier for the Eisenhower government to resist, or rather to deflect, the pressure they heavily felt towards some form of intervention.

The men who planned the 1956 Anglo-French intervention at Suez, which coincided with the Hungarian Rising, had a scenario of their own. Selwyn Lloyd, Foreign Secretary in the Eden government, directed that the United Nations be used to set the stage for the intervention. The metaphor here was his own: those who dramatize a version of politics are often quite conscious of what they are doing.

The British and French presented a resolution to the Security Council which they knew the Soviet Union would veto, and which the Soviet Union did in fact veto. The versatility of this form of drama is shown by the fact that the very same situation—"Security Council Paralyzed by the Veto"—which served Eisenhower and Dulles as a pretext for not intervening in Hungary, served Eden and Mollet as a preparation *for intervention* in Egypt on the theory that, having exhausted the recourse to the procedures of the Charter and having been denied peaceful redress, they had recovered their freedom of action. It was most unfortunate for them that their scenario clashed with the scenario selected by a greater power. The United States, which would certainly have used some equivalent of the Anglo-French scenario if it had decided on intervention in Hungary, necessarily found that scenario altogether unacceptable when, having decided on nonintervention, it found it convenient to claim that its hands were tied by its commitment to the Charter, and by the Soviet veto. Whatever Eisenhower and Dulles might in other circumstances have decided to do, or not to do, about the invasion of Egypt, their own self-scripted Hungarian role, in the autumn of 1956, made it mandatory for them to discredit the Anglo-French scenario, since, if its credibility could be sustained, it would discredit their own. In my belief, this competition of scenarios and roles, within the same theater, accounts in a considerable degree for the vigor and promptitude with which the U.S. Government brought the Anglo-French performance to its inglorious end. Another performance was substituted to a Canadian-American script. The English and French, to mask their collusion with Israel, both pretended that their invasion of Egypt had been designed to "separate the combatants." The United Nations was now used to simulate the fulfillment of this spurious task. The Security Council was still "paralyzed by the veto," of course, but for dramatic purposes the General Assembly will normally do almost as well, and for some purposes better. In this case the General Assembly recommended the setting up of a United Nations Expeditionary Force. Some member nations, encouraged by the United States, constituted a small force which was itself essentially of a symbolic and dramatic rather than military character. Egypt agreed to accept this force on her soil, provided the British and French withdrew. The British and French yielded, in substance, to American pressure to withdraw, but presented their withdrawal in the United Nations theater, as an act of conformity with the United Nations Charter. All they had ever wanted, they said —in an unexpected and unhappy twist of their old scenario —was that the United Nations itself should "act." (They meant to be understood as having wanted it to "act" in the real-life sense although, as we know, they had really only wanted it to act as on a stage.) Now, as a result of their intervention, the United Nations *had* acted and therefore Britain and France, as good United Nations members, were ready, at the request of the United Nations, to withdraw their troops. Unlike the Russians in Hungary, British delegates were seen, by large television audiences in their own country, playing active and virtuous roles in the dramatization of this version of history. This scenario also, in strict logic, had possibilities of collision with the prevailing American scenario, but in practice it was not taken seriously outside Britain itself, and not by any means by everybody there. But it did provide an impressive setting, and as dignified roles as possible, for statesmen immersed in substantive humiliation. The availability of these roles may have helped those concerned to avoid perseverance in their gamble.

The Western powers and in particular the United States have made the most use of the theatrical possibilities of the United Nations; no doubt because they have the greatest say in the management of the theater in question. But others have used it also. The Soviet Union used it, for example, in the Six-Day War crisis in a manner closely analogous to the compensatory scenes of the American scenario over Hungary. Their rhetoric committed them to some kind of support for the Arabs; they convened the General Assembly in order to provide that support in theatrical rather than physical form. Again, in the Cuban missile crisis, Khrushchev used the United Nations in much the same way as Eden had had to use it over Suez: he presented what was in fact a climb-down before America's power, as law-abiding acquiesence with a United Nations request, and he was seen by his own audience playing out a virtuous part.

None of these performances was wholly satisfactory, but each of them was greatly preferable, from the point of view of those involved, to the ignominy of the naked truth. And in most cases the protection given to those involved was also a form of protection for those who watched, and whose danger might have been greater if a stage and suitable roles had not been available for acting out a version of history different from that of the cabinet rooms and chanceries. Nor was it entirely a case of deceiving the public. If the public was deceived, it was often that it wished to be deceived. A wide public, which included perhaps most of us, wished to be honorably indignant about the Russian suppression of Hungary, but without running the risk of nuclear war. The miming of collective paralysis at the United Nations met the needs of this dilemma, much as somewhat similar symptoms in individuals are a way of responding to more directly personal emotional dilemmas. In both cases, we can plausibly infer the coming into play of a kind of survival mechanism. Yet mechanism is an inadequate word to convey the resourcefulness and spontaneity and versatility with which human beings—including such apparently relatively unimaginative human beings as Dulles and Eden—can respond to a simultaneous forked threat to survival and to dignity, by improvising in so many different ways a choreography of dignified retreat and evasion. The metaphor of art form fits more closely than the metaphor of mechanism. Art, as Edmund Burke said, is man's nature. I shall return to that thought.

The boundary between what we call "art" and what we

call "real life" is an illusion. Man's instinct for self-preservation, and his need for dignity, force him to be a theatrical animal. Since he could not survive without expedients which are often in fact undignified, he feels, and satisfies, the need for further expedients, dignifying the undignified. These expedients can be found in historiography, sermons, journalism, and popular balladry, but best of all in the dramatic form of the heroic theater—the most vivid of all ways of changing history with dignity. Man's dignity is his own invention, and life constantly forces him to reinvent it. The exaltation of Irish history in *Cathleen Ni Houlihan* brought a death sentence to Constance Markiewicz, as a corresponding exaltation among black people in this country—expressed in the drama of Le Roi Jones and others—may well have sent out:

Certain men the white man shot.

Tragedy may purge of pity and terror, perhaps. It certainly does not purge away, but rather strengthens, man's sense of dignity, if necessary, before survival. Comedy, on the other hand, mostly prefers survival to dignity. Laughter is an acceptance of the idea that life, even with loss of dignity, remains worth living. The curious dramatic forms of "real life" politics are often intermediate between comedy and tragedy. In conditions where peace is saved, the plot is near tragedy; that is, it had to come as near to tragedy as is felt to be compatible with survival. The demeanor of the participants is in the tragic mode, solemn and heroic; any injection of overt comedy would wreck the whole scenario. Yet the spirit at work is the comic spirit, preferring survival to dignity, but offering the public a simulation of dignity, which it knows will be gratefully accepted as the real thing. Sancho Panza plays Quixote, and tragedy is averted.

Truth is said to be the first casualty in time of war. But it is just as true to say that truth is a casualty whenever the peace is saved. In fact, literal truth is inadequate to human needs, either in peace or war. Man cannot live on ascertained facts alone; he has to make something up: a poem, a religion, a country, an ideology, something you can live with and live for. Curiosity, the need to find out the truth, is part of man's nature, but the truth may be too much to take, and the need to invent, to make up a story, is part of his nature too.

As man's capacity for self-destruction enormously increases, and as the vast increase in his numbers presents new and pressing threats to his individual dignity, so there are proportionate new demands on his ingenuity to find new ways both of surviving and of preserving dignity. The peace protest movements, from the Campaign for Nuclear Disarmament to the antiwar demonstrations, were one set of responses to this dual need. These movements proceeded, of course, by essentially dramatic methods: acting out, confrontation, demonstration. The history of some participants in these movements reveals the tension in the interior of the dual demand for human dignity and for survival. Some who joined the peace movement out of a passion for human dignity moved away from pacifism to a cult of violence. Some workers for civil rights, for blacks in America or Catholics in Northern Ireland, having won considerable support and success by the use of nonviolent methods in the teeth of violence by their adversaries, then swung round and adopted some variant of their adversaries' methods, thereby losing much of their original support.

Dramatic politics, the politics of stylized confrontations, requires an audience that is itself split, or capable of being split; or actually more than one audience. Thus, the activities of the civil rights movement in Dixie and the spectacle of the reprisals which they endured worked by appealing to an audience in the North. Similar activities in South Africa were a total failure because there was no equivalent of the American northern audience. Similarly, the civil rights movement in Northern Ireland scored successes by appealing in effect to a *British* public, over the heads of the local Protestant rulers of Northern Ireland, in the same way in which the American civil rights movement had appealed to northern whites, over the heads of the white supremacists of Dixie. But the renewed physical force of the Catholic Republicans in Northern Ireland and of, say, the Black Panthers, repelled many of those through whose support the civil rights movement won its successes.

Dramatic politics is perhaps most effective when it works at an instinctive level, with that improvising cunning and audacity which man can draw on in a tight corner. It is when self-consciousness comes in that the spectacle begins to become ineffective, the actors losing their sense of where their audience is and where their hopes lie. Nostalgia for one's own past roles, and emulation of the past roles of others, can be pitfalls for the real-life actor, as well as the one on the stage. Thus, Anthony Eden destroyed himself politically through a mental scenario in which he had to destroy Hitler, in 1956, Nasser being cast in the role of Hitler. By a similar process, students in the Universities of Nanterre and Vincennes subsequently reenacted the scenes of May 1968, and young boys in Derry throw stones in memory of August 1969. There are decadent phases in dramatized politics, as in other art forms.

Theatrical metaphor pervades journalistic and general discussion of all public activities, including politics and especially high international politics. The expression "to play a part" is so common that we have ceased to be conscious of it as a theatrical metaphor. Expressions like "the chief actors," "the leading role," "the world stage" have lost significance by their sheer triteness. The word "role" has passed into a more specialized and relatively precise vocabulary, that of the social scientists, but again, I think, without retaining much metaphorical life, or much feeling of its theatrical origin. The significance of the violence and persistence of theatrical metaphor, as a system of referring to public activities of all kinds, has, I think, been largely overlooked, and some very suggestive comments by the late Thurman Arnold and Johann Huizinga and some others have not been followed up. There are several reasons why this may be so.

One reason is the increase in academic specialization, with the restriction of the once extensive area of the old literary culture. Those who study literature generally are, and are generally encouraged to be, both ill-informed and unsuspecting about the working and playing of the society in which they live. Those who have most to say about metaphor, therefore, tend to discuss it within a special and discrete canon, and without much awareness of the difficulty in drawing distinctions between the literal and the metaphorical, in the working of institutions and in the way in which we form pictures of, among other things, political life. Social scientists, on the other hand, often seem relatively ill-informed and unsuspecting about the work and play of language, overconfident about the precision of literalness, and unconscious of the suppressed and often perverted poetry which wells up in their own prose. Yet the language available for their use is the most significant product of the society they study, and a method of studying the society that is insensitive and indifferent to its language will be less illuminating than it ought to be.

A second reason for not studying dramatized politics is political. Politically committed people will readily discern political dramatization as being among the resources with which adversaries deceive the public; they will often—not always—be reluctant to think of the theatrical as one of the dimensions of all politics, including their own. Governments and establishments have strong reasons for not favoring investigation into the political show business from which they have, by definition, profited. The left, on the other hand, which one might think had an equivalent interest in dismantling at least establishment show business, has not shown anything like as much interest in this area as its importance seems to require. This lack of interest may perhaps be traced to two type-figures which have always existed on the left: the puritan and the pedant. The puritan dislikes theater in itself, and is unwilling to look at it, even critically. He finds it repulsively frivolous to see political life itself as theatrical, since politics is important and the theater is not. The pedant, on the other hand, disdains all the surfaces of life; these are for the vulgar; deep underlying causes are the thing for him. He is like a man who is so preoccupied with what he thinks may be going on behind the scenes that he fails to notice what is actually happening on the stage, and how it affects the audience.

But apart from academic and political considerations, I suspect there may be more fundamental human reasons for not looking too seriously into this question. Our dignity is involved, not just the dignity of a few, but the dignity of all of us. There is a strong religious side to political drama and dramatized politics. It was not for nothing that one spectator called *Cathleen Ni Houlihan* "a sort of gospel" and another "a sort of sacrament." This is not a frame of mind conducive or receptive to analysis or criticism. In the case of dramatized politics also there is, especially in moments of great danger, a strong wish to believe in the spectacle as real. Identification with a national leader or spokesman represses any intimation that he may be playing a part in the theatrical sense. Even when it was proved—as it was for example in the case of Adlai Stevenson in the Bay of Pigs crisis—that an actor's lines belonged strictly to the world of fiction, a wide public wished to forget this, and resented, and still resents, any reference to the fact, very much as a religious public may resent any reference to the exposure of a bogus miracle. It is not that the public any longer argues for the reality of the miracle; it is just that references to it wound the public in its essential piety, in its will to believe, and in its dignity.

It is also true that the working of such scenarios as I have described, not only in the United Nations but in the theater of domestic politics and in a wide variety of public transactions, depends on agreement to accept the scenario literally and on its own terms. Illusion—some degree of illusion at least—is a necessary part of how the thing really works; both the preservation of dignity and even survival may at times depend on illusion. Yet it would be wrong, I think, to conclude that because of this we should refrain from analysis of such scenarios, from trying to find the limits of literalness, the social functions of metaphor, and the versatile ambiguous feats of man the dramatist. It is in man's nature to pretend, but it is also in his nature to find out. There are gray areas, too: sometimes man pretends to find out, sometimes he finds out more than he pretends. The acquiescence in certain political spectacles may perhaps fall short of literal credence; it may not be so very different from that suspension of disbelief which we are thought to accord the theater. It may even be that, as against all the resistances I have mentioned, there is now developing, for survival's sake, a greater wish to understand how these matters work. The politics of confrontation, sketchy though it usually is, has forced the dramatic element in politics into the public consciousness. If we are to succeed in living together, in what will for a long time be increasingly and frighteningly crowded conditions, we shall need to develop gestures that require no great elbowroom. It is not so much that there will be a greater need for the symbolic element in politics, but that this element will need to become more economic and more refined. This will require, I believe, in the public generally, a more conscious *general* recognition of the symbolic element, certainly to take it seriously, but no longer quite so literally—a development analogous to the Reformation. There are some signs that this may already be happening in Japan.

More narrowly, among students of human behavior, there should, I think, be a growing recognition that art, as well as science, has light to throw on these matters and that the practice of forms of art, and of the dramatic form in particular, is by no means confined to the small category of those whom we usually classify as artists. Yeats suspected that even political passion was "no more than the desire to be an artist." There is a certain artistic arrogance in that formulation, but it contains a sound intuition, analogous to Burke's about art being man's nature. Burke, of course, used the word "art" in a wider sense than we use it today: his sense would include what we today would call technology, but he did not mean that "technology is a man's nature." He meant, I think, that it was in man's nature to be ingeniously creative, flowering in scientific achievement and in political institutions, but also in art in our modern sense. The fact that I have had to use twenty words to say what Burke said in four—and not merely through my own ineptitude or prolixity but because Burke's saying is no longer readily intelligible—this small fact is a clue to what has been lost in a process of specialization, over a period of less than two centuries. Because, if what Burke says about man is true, and if we have lost the ability to say it briefly, pithily, and intelligibly, then we are further from a sense of our identity than we were in his time. He foresaw that we might be. But we came that way through the very working of the art that is our nature.

The challenges of the period we are now entering are making man devote more of his dangerous "art" to the study of his dangerous nature. It may be, of course, in our nature to deceive ourselves. The artist, as Nietzsche said, is a liar. We do not need, however, to understand ourselves *fully*; perhaps we need, rather, *not* to do so. We just need to understand ourselves well enough to make possible survival, with a certain minimum of dignity for the survivors. Such a condition seems within reach, although how much pain and destruction will be incurred on the path to its attainment we cannot know. But it seems possible that by a fuller understanding of the odd devices by which we act out in public a version of our problems and our natural woes, we might be able to make that path a little less difficult and dangerous. It is with that hope in mind that I have invited you, with your wealth of varied experience and skills, to consider the significance of actors, roles, and stages in an unceasing human tragedy and comedy.

Diversity

The New Exodus
Vine Deloria, Jr.
National Congress of American Indians

Attorney and author. A Standing Rock Sioux, active in social action programs on behalf of Indians in America and keen student of ethnic minorities and their interests in American society. Former Executive Director, National Congress of American Indians. Author of Custer Died for Your Sins *(1969),* We Talk, You Listen *(1970),* God Is Red *(1973).*

At the height of the civil rights movement in 1966 the National Congress of American Indians printed a little business card with the simple inscription "We Shall Overrun" on it. It took a certain amount of courage in those days to suggest that perhaps we would not all integrate immediately upon crossing the bridge at Selma, Alabama. Even more, the thought that American society was on the verge of breaking into a number of competing and conflicting ethnic groups struggling for a new sense of identity had hardly been suggested. Yet the past few years have shown such a startling reversal of the integrationist philosophy that many minority-group spokesmen vehemently deny ever supporting integration.

The rapid change came about, of course, by the promulgation of the concept of power and its acceptance and popularization by young people within the respective minority communities. Martin Luther King and the Southern Christian Leadership Conference used the technique of street demonstrations as a symbolic method of conveying the constitutional issue of equality before the law to a society that understood little of legal issue but could instantly involve itself, via the communications media, with the bombing of Sunday schools and the murder of innocent people. Integration was cloaked in negative terms and the desire of the large majority of the American people was to fulfill the promises made nearly a century before of full citizenship for the American black community.

Not so with the concept of power, however, since power spoke of an aggressive movement to push ahead into unknown areas in racial relations. Power spoke of the ability, necessity, and desirability of groups to exclude nongroup members to preserve their own integrity. Thus social movement became a matter of addressing two communities that were as separate and distinct as two communities had ever been. The fight for equality had essentially been one of the excluded black community addressing the white community with the plea that, should equality be granted, the black community would dissolve itself and become one with the white community. Inherent in this ideology was the shame unconsciously accepted by the divergent minority groups of the differences which they recognized between themselves and the Anglo-Saxon-oriented white majority. Integration was group suicide in the truest sense of the term.

The concept of power split the audience of observers into two groups—the white majority, from whom the particular minority group was declaring its independence, and the minority community itself, which badly needed to define its new relationship both to itself and its former white allies. When this dilemma was not seen in its fullest logical implications, and confrontation tactics continued to be the symbolism by which ideologies were expressed, the social movement split into a number of defensive groupings that appeared to be aligned according to economic status rather than racial or ethnic backgrounds. In turn, this development appeared to threaten the economic interests to which each segment of American society had previously sought to relate. The backlash of the 1968 elections and the subsequent effort on the part of the Nixon administration to roll back the years of civil rights struggle by packing the Supreme Court with southern conservatives were a natural reaction of the white community that had suddenly been thrown into a defensive position for the first time in its American experience.

Then, after two humiliating defeats in the Senate and the determination of the federal court system that the constitutional guarantees of equality before the law be enforced, the Nixon administration began its interest in enforcing the desegregation decrees against southern schools. Thus we witnessed, in a mere four years, turns of 180 degrees by representatives of what had been the old right and left only a decade and a half before. While the disorientation was tremendous, it was overshadowed by the task of finding our way out of the wilderness of social conflict—if not to the Promised Land of equal opportunity, at least to a plateau upon which we could rest and gather our bearings before embarking on our tedious journey forward.

In short, we have now left the comfortable land of assimilation and have been thrust into the outer darkness of ethnicity, while every tool that we have to gather information to find our way was designed for a world of assimilation and integration. Our government, our economic system, our educational system, and the basic documents of our society are built on other premises than those which we are coming to recognize today. Our Constitution is built upon the integration of faceless individuals who band together in search of law and order for the protection of their property. It neither recognizes the uniqueness of a man's social and cultural background nor the changes which take place when that individual is asked to place his allegiance in a specific group of people, all sharing with him peculiar and unique characteristics, language, world views, and, in many cases, economic impotence.

Again, in the field of religion, we have left the comfortable shores of the Christian faith with its absolute formulas of faith and salvation, and we have confronted—almost as if we stood once again at some remote stage of prehistory—the divine and unrevealed reality beyond us. So intertwined had Christianity and our political system become that it was not until June 1970 that our court system defined conscientious objection in any other moral framework than traditional Christian theology tempered somewhat to the movements of the times.

With this burden, almost unbearable in itself, the release of sexual tensions in a society that looked at anything more serious than mouth-to-mouth resuscitation as an unnatural act, the casual approach to sexuality by the younger generation in recent years, coupled with the propensity of their parents to purchase books on attaining orgasms—as reflected by the fact that many of the best sellers in the nonfiction field have been inspirational tracts on copulation—has forced a new lifestyle into a confrontation for which we are not prepared and do not understand. The alleged sexual promiscuity which had been the genius or shortcoming (depending on one's point of view) of the minority groups has now become a casual part of American life.

Meeting in the arena of social movements at the end of the sixties, shouts of repression and revolution hammered at us from all sides. Excluding the fringes who were categorized as communist or fascist, depending upon one's point of view, we found a frightened, confused majority of American society plagued with a series of Presidential Reports, which authoritatively confirmed everyone's worst fears.

The very media by which we receive our information concerning the world continue to be attacked, and credibility has become the watchword when any pronouncement is made on any subject.

I would suggest, nevertheless, that we are in much better shape than anyone could hope to realize. We may be entering a new era of the most significant type of civilization the world has ever seen. But we had better give some serious thought to the realities that face us. We must give up the cherished myths that have blinded us to the reality of American history. And we must be willing to accept the cosmopolitan existence which our technology has given us or we may not see the end of the decade.

For nearly two centuries we have blithely accepted the idea of the melting pot as the epitome of civilized existence but we have acted exactly opposite to our alleged beliefs. Almost from the start, settlement on this continent has been the systematic invasion by groups of people from particular European nations, augmented at times by Africans and Asians who provided the needed slave labor to develop the great economic movements of land settlement and railroad building, development of the continental landmass, and construction of the distribution system by which the economic exploitation of the continent could be achieved.

Early settlements emphasized the determination of the immigrants to build for themselves upon this continent a new version of their ancient homelands. Thus New England, New France, New Spain, New Netherlands, and New Sweden characterized the types of settlements on Atlantic shores that provided the nucleus from which the United States developed. We often say that whites came to this continent to escape various forms of oppression, religious, economic, and cultural. With the development of "new" forms of the old country by people so frustrated with the societies of their homelands that they were willing to risk an unknown land filled with extreme dangers and, in their eyes, savages, we may come to a fundamental understanding of the nature of western European immigration. If the western hemisphere had not existed as a refuge and escape valve for Europe, there would have been a series of revolutions in the European countries of unparalleled and unimagined intensity, instead of the rather mild and sedate revolutions of 1688 in England, 1789 in France, the 1840s in Italy, and the later developments in Germany. Where there was no escape valve of significant immigration, as in Russia, the lid blew off in 1917—with drastic results for the world ever since.

If, then, we understand that the desire for change, opportunity, and identity moving through the societies of western European nations was transferred to this continent intact, we can understand a number of things about American history and our present situation that have not been immediately apparent. Our fierce competitiveness, for example, may well be a negative value, a desperate thrashing about in search of self rather than an expression of manhood and Christian virtue. Our concern with celebrities and glamor may be compensatory for the royalty that we never allowed, for fear of falling into the European past and, in effect, betraying the revolution that had been created by immigration.

The important point in examining American society in this light is that minority groups have never been involved in this process in any significant way. Black athletes were able to intrude into the process at certain points and Indian chiefs have always had both the aspect of celebrity and royalty, but in other than such isolated instances minority groups have been shut out of the social marathon almost completely. They have, in fact, been treated like minority groups, like groups so different in kind and with such twisted potential that they should not even be allowed to start the race, let alone be accorded the opportunity to finish the race.

But then, never have any other groups really been accorded this opportunity. From the start there has been continual differentiation between groups, and this phenomenon has been adequately discussed by countless expositors of the American faith. This has appeared to have been the history of groups in America—disappearance into the melting pot. But, in fact, it has not happened. The opposite has been true. Elections in the nation are carefully built upon piling the correct coalitions of ethnic voters into the proper states and cities to compile the necessary electoral votes to win. In 1966 Lee Metcalf, in a tight Senate race in Montana, remarked that not only was the Indian vote important to him but that the outcome could hinge on the Yugoslavian vote in that state.

Instead, therefore, of successive groups merging into an anomalous mass, there have been a succession of conquests by different national ethnic groups within the American political system. Because each group was western European, it appeared that little change had taken place each time a group moved into a position of power. It was only when the black community pushed into the area of voting rights that a distinction was made. So fast did things move in the last years of the sixties that the American Indian and Mexican-American communities did not have an opportunity either to get organized or to establish a tradition of political combat as a group. They were both massed together in Robert Kennedy's temporary primary coalitions, but this was a coattail effect; they were not fully participating members of an ethnic political coalition.

At the present time there is no national policy toward

minority groups other than the temporary political maneuvering designed to placate southern voters or to further damage the old liberal-integrationist New Deal coalition. The blame does not lie wholly with the Nixon administration. Its political moves have been too blatant and inept to have created the present situation by itself. And many members of the administration have sought some type of answer to the problems of minority groups.

In large part, the present impasse in social movements comes from an inability of any spokesmen of minority groups to relate movements within their group to comparable movements on the contemporary scene or in a historical perspective. Thus, the field of race relations is adrift without a technical vocabulary by which it can affix itself to the experiences of a majority of the American populace. The very words, it appears, can be used and have been used so often and in so many contexts that they make bad publicity releases and communicate little besides the hopeless stereotypes of a bygone era.

One example is the concept of black capitalism. As put forth by the Congress of Racial Equality in the proposed Community Self-Determination Act introduced in Congress in 1968, black capitalism was a plan by which ownership of community facilities, including employment and schools, could be created within various black neighborhoods. But the use of the term was unfortunate. Shortly after its introduction people derided black capitalism as an effort to make a few black millionaires at the expense of the poverty-stricken millions in the ghettos. The same historical experience of the white community with its robber-baron past was foisted upon the conceptual innovation of black capitalism to interpret the term. It was rejected by many in the New Left as a compromise with the capitalism that had brought the nation to its present disastrous situation.

Here again, the essence of the CORE idea never reached a significant number of people, so that its easy label and sloganesque neatness doomed it from the start. In rejecting black capitalism, vocal activists called for socialism, and Huey Newton announced that civil rights laws produced no change for blacks and called for a socialist takeover. Such a declaration overlooked the fundamental premise of black power and community self-determination upon which it was built. Socialism, in the context of self-awareness of racial and ethnic groups, is not a viable alternative to capitalism. If anything, it is more stifling, more integrationist-oriented than capitalism. Whether Newton advocated socialism as an economic alternative or as a political alternative to what we have is difficult to understand. The important thing to note is that the message, ill-defined as it was, served primarily to increase the fears of the white community and its component ethnic groups that the movement for independence by the black community was directly opposed to its vested interests.

In the capitalism-socialism choice little analysis has been made by leadership in the minority groups as to the actual situation we face. Few industries survive today without massive government transfusions in the forms of grants, contracts, and subsidies. Whole industries, such as aircraft manufacturers and ship builders, are in fact government corporations thinly disguised as independent economic entities after the classic capitalist model. States are heavily dependent on government funds for most of their social welfare programs, educational institutions are a treadmill of grantsmanship. Even the few regional development plans, such as the Tennessee Valley and Missouri Valley Authorities; the Four Corners Development program in Colorado, Arizona, New Mexico, and Utah; and the major economic groupings, such as the Federal Reserve System, are built on suprastate entities, organizations, and structures that span a number of states to develop resources that one or more states could not develop alone.

While the original conception of the federal system was the coalition of sovereign states, operating with a minimum of interference from the national government, that concept has all but vanished. In Colorado, former Governor John Love once discussed splitting his state into a number of regional capitals to handle the particular problems of particular groups and industries. The boundaries of the states were originally artificial lines drawn across the map by long-forgotten political deals, and were influenced, perhaps, more by considerations of slavery and the economy than by any determination to create a society self-sufficient in itself. One need only to look at the states of Washington, Idaho, and Montana on a topographic map to realize the virtual impossibility of creating a community of self-interest on the present political structure when the land itself mitigates against such a development.

Minority groups have not faced the artificiality of the present structure, nor have they brought it to the attention of the white populace. Instead, they have largely reacted against the structure as if it were set in a permanence that would forever remain beyond the possibility of change. In so doing they have largely accepted the mythology of early colonial America without challenging any of the basic philosophical concepts of those ancient times. A call for revolution, therefore, has a hollow ring, because it does not take into account the unconscious movements of history, which by and large refused to recognize the boundaries and the concepts that were objectively articulated, and created developments built, in fact, around the land and resources.

Within our present political framework there is but one political status which has remained in a supraconstitutional condition. That is the undefined and perhaps undefinable position of the Indian tribe. Political doctrines of the status of the Indian tribe vary radically from one extreme to another. A tribe is at once an entity "higher than a state" (Native American Church v. Navajo Tribal Council) and an abject ward of the government for whom Congress has little but contempt and with whom Congress can be entirely arbitrary, due process and equal protection or not (Lone Wolf v. Hitchcock).

Comparison of the Indian tribe with TVA, MVA, and the regional offices of the various government departments shows the startling parallels in political status, economic viability, and utter dependence on the tide of public opinion. The fortunes of all of these entities vary according to the philosophy of the political party occupying the White House. An economy cut with any severity of purpose will curtail developments of all of these organizations with telling effect. In this sense they differ from the traditional groupings of political structure aligned according to state and county governments. And they are more responsive to change. TVA, for example, is more politically sensitive to the wishes of its constituency than has been the Tennessee legislature, which had to be virtually attacked by the Supreme Court in Baker v. Carr in order to force it to reform its apportionment.

The difference between an Indian tribe and a regional de-

Oscar Bear Runner stands guard as a teepee is set up for official negotiations at Wounded Knee, South Dakota, in March 1973. Courtesy of Wide World Photos.

velopment authority has been in the refusal of the federal court system to make some important decisions regarding the former. Thus, a difficult question with respect to Indian rights is often called a political question, and the courts have refused to speak on the subject. With the regional development agencies, courts have generally called difficult questions "administrative decisions," and have refused to overrule what the bureaucratic structure has decided. Since the structure at its apex is political, and the top jobs change by political ap-

Indians keep watch over the countryside at Wounded Knee, South Dakota, in December 1973. Courtesy of Wide World Photos.

pointment, the difference is, in fact, one of language and not of substance.

Again, the minority groups have failed to see that the formal structure operates in much the same way as does the private arena. Certainly the decision to placate the South by the Nixon administration was one of political motivation, although it did not surface within the governing structure but was one of political significance in Congress. Even in the area of job placement, educational opportunity, development of state programs, and social welfare programs, decisions were primarily political, although without the formal recognized status received by regional development corporations and Indian tribes. Yet, formally or informally, recognized or unrecognized, the *process* by which we relate various segments of American society and political structure is much more important than are the names to which we commit ourselves when describing the process. We have allowed the passionate desire of the western European to label things to carry us beyond our ability to comprehend the manner in which the things we have labeled relate.

Identifying the American process as essentially political in its dynamics has, of course, been no novel discovery, but here we have generally stopped. When the tickets were balanced according to ethnic group, when we had picked our Polish Postmaster, Anglo-Saxon attorney general, and filled our Jewish seat on the Supreme Court, we retired from the field satisfied that the coalition would hold up through both terms of our winning candidate. And while groups have been informally identified by their places on the ticket, there has been no frank acknowledgment of what has happened. Consequently, the necessity for defining the relationships that will continue to exist between groups during the ensuing administration has been neglected. While each group has token representation, no effort has been made to relate the strengths of the respective groups one to another. It is impossible under our political system to do so. The system is neutral and does not formally recognize what has informally taken place.

At this point the minority groups and their liberal allies have missed an important point which southern conservatives have been quick to note. The civil rights movement, with its ensuing legislation, was ostensibly a movement to guarantee to all Americans certain inalienable rights. In fact it was class legislation or group legislation designed to accelerate the economic development of a certain group—the black community—and to provide for them certain safeguards by which they could achieve a status within the American system to enjoy that development.

Almost coincident with the rise of the power movements came a flowering of ethnic developments of tremendous variety. People of German descent had an Oktober Fest, the Scandinavians celebrated Leif Ericson Day, the Italian Americans had a protest against harassment of some of their more celebrated citizens by the FBI. Unconsciously or consciously, segments within the white community had recognized the necessity to assert their group sovereignty as a defensive measure against the successes of the black community. Coming with the apparent racial backlash, the movement toward group integrity was largely overlooked or casually passed off as a normal and periodic assertion of the uniqueness of what each group "contributed" to American society.

The present scene is, therefore, characterized more by the reawakenings of numerous groups of particular peoples than by the widening split between black and white envisioned by the Kerner Report. Assertion of the black community that it had power and integrity unto itself and would be greatly weakened by integration in fact freed the white man from his burden of maintaining an undefinable status as "American" and allowed him to prefix his Americanism by his particular European national background. The denial of specific identity, the denial of community, which each immigrant had suffered on his entry into the American political system and which he had partially regained at election time, was ended. In a real sense a substantial number of citizens had left the desert of despair, while an even greater number had been forced into the same desert.

Developments of the past several years have given specific and startling examples of how this unheralded movement to ethnicity has affected us. Sirhan Sirhan, although to all intents and purposes an assimilated American, became so upset over the possibility of Robert Kennedy supporting Israel that he was driven to assassination to prevent Kennedy from becoming President. The Kennedys themselves serve as an example of Irish ethnicism, influential on a world scale. The travels of Jimmy Breslin and Pete Hamill to Ireland and their intimate acquaintance with Irish political events only serve to further highlight the situation.

Since World War II the United States has taken on the role of world policeman and become the influential world power. Unable to solve its internal inconsistency, the United States is now subject internally to world events and changes, as the various nations group and regroup in world coalitions and these groupings are reflected by realignments on the domestic scene. Further involvement in world affairs will result in increasing tensions domestically as the United States appears to favor one side or another in overseas disputes.

Our legal system, as it has sought to regulate the relationships between individuals, has been based upon the right of two individuals to contract one with another. Any infringement of that contract or the contracting power has been stricken down as an action irreconcilable with our Constitution. While the simplistic contracting relationship may continue to hold between individuals, it cannot hold between groups. Groups have a supraindividual aspect to them that speaks of nationhood—and nations sign treaties, not contracts. In this, they differ: contracts define the letter of the agreement, while treaties define the general moral relationship between peoples. Only the individual signing a contract may break his own contract, but certain members of a group can break the treaty of a group even though the vast majority of that group faithfully maintained its part of the agreement. It is the moral integrity of the group itself that maintains a treaty, but it is only apparent intent that contracting parties are held to.

We cannot approach intergroup or interracial problems on a contractual basis. In a sense that was already attempted after the Civil War. A century later, a substantial number of white citizens felt compelled, on moral grounds, to fulfill the promise of citizenship given at that time. For a hundred years the black community had appealed for its rights on a contractual basis, without any significant satisfaction. It was only when the peoplehood of the blacks cried out to the white community that response was possible. The great discovery of the civil rights movement was that blacks continued to be black even though the legislation promised that they would be the same as whites. It was at that point that Carmichael and Hamilton discovered the strength of the black community, which had been denied since the people had arrived from Africa.

If we are to relate group to group, people to people, then we are on the threshold of a new era of our existence. We are like wandering tribes of prehistory, discovering for the first time that other groups of our species live just beyond the mountain. We are like the early colonies, discovering Indian tribes inhabiting the wilderness. We are like Europe, developing from the ruins of the Roman Empire. We can use the American political arena to allow one group to oppress another, or we can use it as a forum, an arena in which the problems of our society, and perhaps the world, can finally be resolved.

The Constitution, as it has come down to us through two centuries of hardship and pain, can be the tested ground rules of the redefinition of our society according to the uniqueness and integrity of our respective constituent groups. In this sense it shows every promise of being comparable to the laws of Moses, delivered in the wilderness to the tribes of Israel, which provided the framework for Hebrew-Jewish existence for 5,000 years. In that context the smallest tribe was equal to the largest and most powerful. In the present context Indian treaties—the only existing documents by which two peoples are related within the American constitutional framework—are regarded as the weathervane of trends. These treaties require a willingness to go beyond words and formulas to the intent of the two treaty-making parties—to live in peace and friendship forever.

There is no force except a sense of morality and integrity that can make the American government keep its Indian treaties. By the same token no promise, implied or explicit, made to any other group can really be enforced or fulfilled except by the willingness of people to keep the promises their hearts made, and not to rely solely upon the words their lips uttered or their hands put on paper.

With the reshaping of American society along ethnic lines, it becomes apparent that our religious understandings must change also. The most potent force for change in the black community is probably the Black Muslim movement, the most influential movement in Indian country is the reawakening of traditional religion. Already our laws are reshaping themselves along a more intangible value system than western Christianity has previously been willing to admit. Good Samaritan laws are replacing the old liability precedents which hampered us from helping our neighbors who had fallen among thieves.

It was a paraphrase of the Exodus situation that dominated Martin Luther King's speech before his death. It is this exact situation that confronts us on the domestic scene. The Woodstock Nation purported to have broken through into new lands. The future will tell. At least the youth have declared their independence. What we see in the United States, we see in most of the industrial countries of the world. At a certain point in technological development the nation outgrows in many ways its original premises and political orientations. Russian intellectuals foresee the breakup of the Soviet monolith within twenty years. The Irish struggle continues.

The United States, Russia, and England attempted to define man in exclusive ways. Russia conceived of man as an economic animal, and through Communism sought to bring him the Promised Land. But man is not an economic animal. England, through a merger of church and state, sought to define man as a religious animal. But man is not exclusively a religious animal. As the esoteric religions of witchcraft and the Catholic-Protestant conflict in Ireland continue to escalate we shall see startling developments in the British Isles. The United States attempted to define man as a political animal. Man is not primarily political, but has traditionally avoided warfare by political means.

For that reason I believe that we have the best opportunity to solve our problems. Better than any other country on earth. We have defined the political guidelines and ground rules. We have only to expand our vision of what man can be and how he conceives of his immediate group, the group that calls to his real self. Russia cannot solve the problem of man economically without centuries of political experience and religious development. England cannot now provide economic growth and political structures within which she can resolve her problems. But we have a system that is basically neutral. It is up to us to fill it with content in which further economic and religious growth can occur for those groups that have not yet had that opportunity. We can achieve a balance—and with our ecologists telling us that mankind has only a generation left, it would be nice if we could have peace before the end of the world.

Cultural Pluralism, Political Power, and Ethnic Studies*

Murray L. Wax
University of Kansas

Anthropologist and educator. Specialist in intercultural relations, especially on the "world view" of particular communities. Professor at the University of Kansas since 1967. Extensive research in formal education of American Indians. Author of "Tree of Social Knowledge," Psychiatry *(May 1965).*

*Revised slightly from initial presentation at the annual meetings of the American Ethnological Society, April 1972, Montreal.[1]

> *... ethnic groups are categories of ascription and identification by the actors themselves, and thus have the characteristics of organizing interaction between people.* FREDRIK BARTH

Since the emergence of separatist and nationalist movements among the blacks, Chicanos, and American Indians, fresh attention has been given to notions of pluralism, especially in educational contexts. In justifying their demands for courses or programs or schools in black studies, Indian studies, and the like, spokesmen for these ethnic-racial groups will refer to distinctive values, orientations toward life, and styles of interaction as, for example, to "soul food," "black thought," or to American Indian attitudes toward nature or modes of relating to fellow human beings. Clearly what is being referred to by these nationalistic spokesmen are characteristics of what they perceive to be distinctive cultures or subcultures, and they themselves speak of "black culture" or "Indian culture." On the other hand, these proponents of black, Indian, and Chicano studies seldom use the phrase "cultural pluralism," perhaps because they are aware of its considerable history in debates about the status of the white immigrants to North America during the late nineteenth and early twentieth centuries.[2] Many of these spokesmen declare emphatically that the immigrant experience has nothing to do with the contemporary situation of their own people. Nevertheless, the phrase "cultural pluralism" does seem to me to denote the nature of their demands within the educational context, and I shall from time to time employ it within this essay, as proves convenient.

Against these demands for programs of ethnic studies, there has been leveled a variety of significant criticisms. For simplicity, permit me to focus upon the Indian case; much of what I have been observing here can be transposed into the other cases. The criticisms offered of programs of Indian studies in universities (or of special programs or schools for Indian children in the lower grades) assume two forms: first, that if there is a distinctive Indian culture, then it is so irrelevant to modern North America that, were it to be taught in the schools, it would prepare children only to live within an isolated reservation enclave; second, that whatever once was unique to a rich Indian culture has been lost, so that all that is left which purports now to be Indian culture is instead merely "a culture of poverty" — or, in other words, a degenerated or "degraded" culture deriving from oppression. The myths, rituals, customs which once composed a rich and integrated culture have been replaced by an aggregate of heuristic devices for maintaining existence under conditions of oppression and semistarvation.[3]

Both of these criticisms are wrong, and I wish to expose the nature of their limitations in order to resolve important theoretical and educational issues. First, I would like to inquire whether the terms "culture" and "cultural" (as in "cultural pluralism") assist or obfuscate our efforts.

Among the Indian peoples with whom I am personally familiar there are certainly traits that have the strongest linkages to aboriginal Indian cultures. On the most obvious level, many of the Oglala Sioux that I know are speakers of Lakota; many of the Cherokee of northeastern Oklahoma are speakers of their native language. And, while a scholar could adopt the counter-Whorfian attitude that language is a technical instrument having nothing to do with anything else in a culture, I think it equally easy to argue that if something as complex as language has survived, then this constitutes *prima facie* evidence that other aspects of culture likewise have been surviving. It is true, of course, that these languages have changed, and in the cases of Sioux and Cherokee it is true in particular that the most elevated and philosophical vocabularies have fallen into disuse; nonetheless, one cannot evade the fact of some kind of linguistic persistence and therefore of some kind of cultural persistence. I can put the case even more forcibly: when Rosalie H. Wax first became involved with Indian groups, she quickly noticed that whether these were Indians at the Chicago Indian Center, or at a University of Colorado Workshop for Indian college students, or at the Pine Ridge Reservation they all manifested some common patterns of social interaction. Together with Robert K. Thomas she described and analyzed these patterns of interaction in an essay titled "American Indians and White People," and this essay has been found valuable by scores of persons who have had to live or work among Indian peoples. Not only has this essay been several times put in anthologies, it also has been duplicated for all kinds of workshops that brought together Indians and non-Indians. I judge it quite safe to say that this would not have happened unless she were describing a social reality.

On the other hand, even the most naive observer of Indian peoples must grant the enormous amount of change that has occurred during the past several centuries.[4] Particularly is this evident in material technology, where the hunting and gathering peoples of the Great Plains have been transformed into users of automobiles, television, portable radios, and coin laundries. The Oglala Sioux still annually perform a Sun Dance at the Pine Ridge Reservation, but no living Sioux could have experienced what it was like to be a horse nomad on the Plains, living off the bison herds and sheltering himself within a tipi. And it is not only that there have been changes for the Sioux people as a whole. Equally important is the loss of homogeneity.

Where the Sioux of the eighteenth and nineteenth centuries practiced a common set of rituals and shared a common store of mythology, today this small cluster of people includes participants in a number of the religious denominations of North America: Roman Catholic, Episcopalian, Methodist, Pentecostal (Holy Roller), Church of Latter-day Saints (Mormon), as well as the Native American Church. In addition, some Sioux still practice a variety of cults, such as *yuwipi* or *wanblee,* whose shamanistic origins are very old. Likewise, the Sioux are remarkably diverse in levels of formal education, or in types of occupation, or in styles of living. Given this heterogeneity, for an observer to attempt to describe something as being "contemporary Sioux culture" becomes a difficult task. What makes this difficult task even harder is the fact that large numbers of Sioux children are spending substantial amounts of their young lives in boarding schools operated by the federal government or mission churches. Children within those schools may learn many things but they do not learn what their ancestors learned on the Great Plains. Nonetheless, the people who are processed by these schools tend to leave with a very strong feeling that they are Sioux (and Indian), even if in terms of specific cultural traits there is very little that they share with their ancestors.

Given this heterogeneity, it would seem evident that in contemporary North America, being an *Indian* is a *social* and *political* identity and not a *cultural* identity. It is a matter of how an individual identifies himself and is identified by others, and of whom he associates with and in what types of interaction, and for what interests. Identity as an Indian is not a matter of the possession of particular cultural traits. Among contemporary Indian spokesmen and leaders are many persons competent in the technology and apparatus of modern industrial society; there is no one fixated at the material culture of Black Hawk or Geronimo. Just as a person can be identified as a Jew in modern North America and yet not be able to speak Yiddish or Hebrew, nor have any but a superficial acquaintance with the Torah (or sacred law), so can one be an Indian without knowledge of traditional Indian languages or familiarity with traditional Indian customs and rituals. But, if one grants both of these instances, then it follows that the phrase "cultural pluralism" was as dubious in the immigrant case as it is now in the case of the contemporary militant ethnic-racial groups. What is at stake is a form of *social* and *political* pluralism.[5]

Of course, I am not denying that features such as skin coloration, physiognomy, body posture, or accented English may serve as stigmata that classify a person as a depressed subcaste within North America. Indians, Negroes, Puerto Ricans, and Mexican Americans have suffered and continue to suffer severe discrimination; so too did (and sometimes still do) Jews and other ethnic minorities of Europe and the Middle East. In personal contacts with WASPs, certain cultural (or linguistic or paralinguistic) traits became (or remain) markers for treatment as inferiors. But to make a long argument short, I would comment merely that we deal here not with cultures but with symbols of relative power and status.

This is not to deny that for immigrants or modern ethnics "culture," as history or tradition, may have a strong symbolic meaning. In their advocacy of programs of Native American Studies, militant Indian leaders have urged the inclusion of courses in native languages which otherwise are in the process of disappearance. A dispassionate observer may endorse their program while still forecasting that, just as the Scandinavian Lutheran Churches in the United States have had to shift the language of their communion to English, despite the feelings of older parishioners that only the Scandinavian mother tongue was suitable as a language in which to address God, so too will North American Indians see the further erosion of many of their native languages.[6] But, if Scandinavians remain a distinct ethnic bloc, despite the loss of their languages, how much more likely it will be that Indians will likewise remain a noticeable bloc within North American society!

In making the foregoing argument, I am challenging the use of the plural concept "cultures" in relationship to the ethnic-racial minorities of modern North America, and I am suggesting that we recall the relevant intellectual history. For Edward B. Tylor there was but a single human *culture,* which was equivalent to *civilization.* Around the world, the anthropologist perceived traits of this (universal human) culture in the process of diffusion, or occasionally, of invention. Tylor took this concept of culture (or civilization) from the German historians. Those scholars had become conscious of the achievements, not only of the societies of Classical Greece and Rome, but also of India, Egypt, and China; and some historians saw in these societies the distinctive, if transient, manifestations of the spirit of mankind. For some of these scholars each civilization became a stage in the evolution of mankind and human reason.

While anthropologists well know that Franz Boas was a product of the natural science regime of German universities, he was certainly also influenced by that larger intellectual climate, and especially so influenced were such of his students as Alfred Kroeber, Robert H. Lowie, and Paul Radin. These men transferred to the North American continent this comparative vision of human civilization, and in order to make evident their respect for the achievements and values of the Indian peoples, they transformed the term "culture" from the singular to plural. So, each American Indian people was viewed as bearing a distinct culture, and each culture was viewed as having an existence that went from birth to death. While the elements of a particular Indian culture — the cultural traits — might have been assembled from a variety of sources (as Tylor had argued), nonetheless each such culture had an integrity and a unity. To the mature Ruth Benedict a culture had the unity of a pottery vessel — a cup — and like such a vessel it could only be whole or shattered. A better simile for the attitude of Boas and his students was that of evolutionary biology, so that cultures were like species that had evolved but would in time become extinct. It appeared to these scholars that the set of aboriginal American Indian cultures were like fragile biological specimens whose environment had suddenly become hostile so that they were in the process of becoming extinct. They saw the task of anthropology as an obligation to the science of the future — namely, to preserve in the archives the records of these specimens of the human spirit. The cups were being shattered, the species were dying, but like the civilizations of ancient Greece, Rome, India, and Egypt the record of their contributions to human life would be preserved — not in stone, but in the pages of monographs.[7]

Since their time, ethnohistorical researchers have forced us to re-examine their assumptions about Indian cultures. It is now well established that the Indian societies of North America had been engaged in processes of cultural change for generations, and that they had responded positively and creatively to the opportunities afforded by European contact and trade. What Boas and his pupils were encountering were not the last pure specimens of aboriginal Indian cultures, but rather a set of peoples who had during the eighteenth and nineteenth centuries worked out a variety of adaptations to their biological and social environments and who were now striving — against fearful and even lethal odds — to create new adaptations. Those whose grandchildren survived are still Indians, or Native Americans, but they no longer practice exactly the same ways of their ancestors. This is not strange, considering that very few of us, whether white, black, Indian, or whatever, practice the exact ways of life of our ancestors. The anthropological error did not lie so much in employing the concepts of pattern and culture as heuristic abstractions. Rather it lay in equating the existence of a *people* to the persistence of specific cultural *traits,* and to the accompanying notion that if and when a particular people stops practicing some of the customs of its ancestors, that particular people thereupon becomes extinct or nonexistent and is no longer able to function as a social or political entity.

Whether Indian or non-Indian, those who conceive of Indian *cultures* in this statically pluralistic fashion and, therefore, as becoming extinct (or degenerating into a culture of poverty) are neglecting that which was central to Tylor's vision of culture, namely the place of technology. Those Europeans who established in the Americas what they called "New England" were able to do so only because they incorporated into the basis of their agricultural existence the maize-beans-squash complex that had been developed by the Indians. The history of the non-Indian peoples of North America would be entirely different were it not for such Indian inventions, which were integrated into their fabric of life. Conversely, that which for most North Americans was the stereotype of Indianness — the Navajo with his sheep or the mounted Indian chasing the buffalo — both of these represent adaptations of European technology by Indian peoples. If we can grant these manifestly simple arguments, then we can see that preserving the plural form of the term "culture" in relationship to the ethnic-racial minorities of North America serves more to obfuscate educational policy than to assist it.

Once we discard the misleading conceptualization of "culture" in relationship to special educational programs of ethnic-racial minorities, we are able to plan in a more creative fashion. On the one hand, we are able to reassert the patent educational truth that the educator and the curriculum should begin with what the child already knows and build on his skills and competencies, rather than analyze what he does *not* know in relationship to some idealized version of the middle-class child. Thus, if the child enters school speaking Lakota or Cherokee, this is a facility which should be built upon rather than condemned as either a deficiency or a vice which, if encouraged, might lead him to spend the rest of his life in a reservation enclave. Equally important, everyone involved can take a fresh look at programs of ethnic studies. For the issue here (and I am not the first to say this) is not that of fitting the Indian student to return to an outmoded or deteriorated reservation environment, but that of helping the student to find himself by giving him a notion of who he is and what his ancestors have been. Especially for ethnic-racial groups like Indians, blacks, and Chicanos, who have been the target of sustained rhetoric that has declared their inferiority to the white Protestant, a program of ethnic studies is of great value in helping to redress the balance and assist all parties to a more accurate moral perception of history. Ethnic studies programs also symbolize a respect by the educator of the skills which the child brings to the school from his home and community.

Insofar as we as anthropologists or other social scientists are involved in programs of Indian or other ethnic studies, we should therefore be among the first to deny the simplistic assumption that to be Indian is to be a hunter-and-gatherer wandering about with feathers in his hair. Correspondingly, we may at appropriate times have to deny the mythology of the special relationship between Indians and nature, inasmuch as the internal combustion engine delivers pollutants to the atmosphere, whether it is located in an urban or a reservation environment.

To assert the notion of socio-political pluralism is to deny the ideology of Indian *assimilation* as it has been set forth ever since the major European invasions of what they saw as the New World. Missionaries, government officials, and reformers believed or hoped that the Indians would either be assimilated into the fabric of European civilized and Christian life, or else would become extinct. In North America during the nineteenth century, missionaries thought that once the Indians were converted to Christianity, taught to speak English, and provided with basic schooling, they would merge into the white population. Some of them did, in fact, so disappear, as one can find in many social gatherings, where individuals who look and act like mainstream Americans will tell you that their grandmother was a niece of Chief Crazy Horse. But it is also true that one can find leaders of contemporary Indian organizations who are as educated and sophisticated as their fellow North Americans and who yet think of themselves and are thought of by others as fully Indian.

Of course, there is another and extremely important dimension to this discussion, and that is *power*, both political and economic (cf. Yetman & Steele, eds. 1971). For the significant characteristic of the enclaved Indian peoples of the late nineteenth century was their loss of political independence and of their economic resources, and their subordination in both respects to organized white or non-Indian interest groups. And, it has been well argued, especially by critics of the past decade, that the distressing fate of the Indian peoples lay not in their lack of integration with non-Indian (white) peoples but rather in the extent and quality of their integration. Similarly, in the case of the Negro in the Deep South of the past century, his difficulties lay not with a lack of integration into the society, but rather in the very nature of his integration into the political and economic life of the times as a depressed and exploited caste.

Accordingly, what is significant about programs of Indian studies or ethnic studies is not merely the ideological content of these programs, but the fact that they represent a transfer into the hands of spokesmen for these groups of a certain quantum of funds and power. White idealists are sometimes sorely troubled when they observe that the ethnic-racial minority in question has been more concerned about how to distribute the patronage represented by these programs — the jobs, the contracts, and the perquisites — than about performing services for the clientele constituted by their students, whether of their own ethnicity or otherwise.[8] But, of course, this activity confirms the theme of this essay that the real issue is not culture or cultural pluralism, but social pluralism and associated political power.

To summarize: contemporary American Indian groups are culturally heterogeneous; some continue the native language or display other traits which have the most direct linkages to the aboriginal culture, while others seem to have blended wholly into one or another version of "mainstream North America" (whether middle, working, or lower class). Despite their cultural heterogeneity, Indians have been organizing to advance their aggregate political interests, whether in rural or urban areas, on tribal, regional, or national bases. Sit-ins, court cases, polemical books, journals, essays, and even newsreels or movies, are part of the armamentarium of these native American political groups. Yet, while Indians have been organizing and acting in these fashions, anthropologists and even Indian leaders themselves have continued to talk as if the distinctive feature of Indians was a common culture, either of a tribal or even a continental character.

I have tried here to cope with this cognitive dissonance by suggesting that it is more appropriate to conceive of Indianness as a socio-political rather than cultural identity. I do not intend to deny the persistence or significance of Indian cultural traits, but only to point out that in contemporary North America, the socio-political membership and activity are the decisive factors. Particularly in the educational context, it is misleading both to Indian and non-Indian students to portray Indianness as if it were a matter of preserving the traits of an aboriginal and static culture.[9]

NOTES

¹ The audience discussion was helpful to me in clarifying my ideas and preparing this revision. In particular, Kathryn T. Molohon directed my attention to the writings of Fredrik Barth on ethnic boundaries (1969); while Joan Cassell suggested that portions of this type of analysis were applicable to the contemporary women's rights movement—a theme which I do not here develop but hope that she will. Some readers may be interested to follow the linkage of the critique of "culture" in this essay with the more extensive discussion in my "Myth and Interrelationship in Social Science" (1969).

² Milton M. Gordon reminds us that the first use of the phrase "cultural pluralism" was by the philosopher Horace M. Kallen, in 1924, when he was explaining the position concerning American multiple-group life that he had developed in essays published during the previous decade. Subsequently, the position, but not the phrase, was developed further by Robert E. Park as he attempted to comprehend his experiences both as a newspaper reporter in metropolitan areas, a secretary to Booker T. Washington, and a graduate student in Germany. Park argued that it was possible for peoples to be organized together either as ecological communities or as cultural societies. As ecological communities, they would be integrated in terms of symbiosis and division of labor, and so would minimize common features and emphasize diversity; while as cultural societies, they would be integrated by a moral consensus. Clearly, the ecological community rests upon "cultural pluralism."

³ While this position with regard to American Indians is a generally popular one among the laity, it has found only a few academic exponents, most notably Bernard James. The more general position regarding "a culture of poverty" was, of course, advanced by Oscar Lewis, and there is by now a rich polemical literature (cf. Leacock, ed., 1971). Rosalie Wax has remarked to me upon the similarity between Lewis's "culture of poverty" and the theory of cultural degeneration which had so many exponents in the early and mid-nineteenth century. In 1810, for example, Samuel Stanhope Smith characterized primitive peoples as "the generative offshoot of an original higher civilization, the product of 'idle' and 'restless spirits' who spurned 'the restraints and subordinations of civil society.'"

⁴ While ethnohistorians as diverse as Eggan, Leacock, Lurie, and Spicer have emphasized the developmental changes among native

American peoples, other authors have continued to present the myth of the unchanging traditional Indian (for a review of this myth see M. Wax, 1968). The loss of traditional Indian homogeneity is a point particularly developed by Spicer (parts III and IV, 1962).

[5] A. W. Singham distinguishes between "the social pluralism of a modern society like America" and "the cultural pluralism of traditional societies": "Whereas in the culturally plural traditional society the very concept of the nation is at stake, in modern societies which are socially plural all groups accorded legitimacy by the society are allowed to participate, despite their retention of primary group loyalties that may be ethnic or religious." The distinction drawn by Gordon (chapter 6, 1964) between "cultural pluralism" and "structural pluralism" is also worth noting.

[6] Among large and enclaved Indian groups such as the Navajo Nation, the native language may continue to flourish. Likewise, native languages may be maintained in specifically religious contexts, as perhaps in the case of the Long House religion among the Iroquois. Yet, many other Indian languages are being eroded, without the group involved being likely to disappear as a social unit. Robert Breunig reports that among the Hopi, the more conservative parents wish the school to teach English exclusively, it is the less traditional who fear the loss of Hopi and wish there to be instruction in that language within the schools. Of course, this discussion does not obviate the fact that a common domestic language (such as Spanish for Mexican Americans, Puerto Ricans, and native Spanish Americans) can be an important instrumentality for social solidarity and a badge of personal identity. Nevertheless, it has been my experience on a number of occasions to encounter professionals who speak but little Spanish yet identify strongly with the cause of their people (Chicanos, Puerto Ricans) and even become public spokesmen.

[7] Sol Kimball reminds me that these American anthropologists were mostly—or often—employed by museums, and that much of their fieldwork was devoted to obtaining museum specimens. Thus, they tended to view native peoples in a vocabulary derived from the museological perspective, as entities whose remains ought to be preserved and enshrined. Barth's comment (1969) is appropriate: "Practically all anthropological reasoning rests on the premise that cultural variation is discontinuous: that there are aggregates of people who essentially share a common culture, and interconnected differences that distinguish each such culture from all others. Since culture is

nothing but a way to describe human behaviour, it would follow that there are discrete groups of people, i.e. ethnic units, to correspond to each culture."

[8] Note here the idealistic expectations of anthropologists and liberal friends of Indians concerning the Rough Rock Demonstration School, and note also the attempt by these parties to suppress the unfavorable evaluations of this school, a controversy reported in "Skirmish at Rough Rock" (1970), a set of articles in *School Review* (volume 79, pages 57-140).

[9] Barth (1969) notes critically that in characterizing ethnic groups "the sharing of a common culture is generally given central importance. In my view much can be gained by regarding this very important feature as an implication or result, rather than a primary and definitional characteristic of ethnic group organization."

REFERENCES

BARTH, FREDRIK, editor. *Ethnic Groups and Boundaries.* Boston: Little Brown, 1969.

ERICKSON, DONALD A., and HENRIETTA SCHWARTZ. *Community School at Rough Rock.* Document PB 184571. Springfield, Va.: U.S. Department of Commerce, 1969.

GORDON, MILTON M. "Assimilation in America: Theory and Reality." Daedalus, 1961. (Reprinted Bobbs-Merrill S-407.)

―――――. *Assimilation in American Life.* New York: Oxford University Press, 1964.

KALLEN, HORACE M. *Culture and Democracy in the United States.* New York: Boni & Liveright, 1924.

LEACOCK, ELEANOR B., editor. *The Culture of Poverty: A Critique.* New York: Simon & Schuster, Clarion Book 20846, 1971.

PARK, ROBERT EZRA. "Symbiosis and Socialization: A Frame of Reference for the Study of Society." *American Journal of Sociology,* volume 45; pages 1-25, 1939. Reprinted in *Human Communities: The City and Human Ecology.* EVERETT C. HUGHES, et al., editors. New York: Free Press, 1952.

SPICER, EDWARD H. *Cycles of Conquest: The Impact of Spain, Mexico, and the United States on the Indians of the Southwest, 1533-1960.* Tucson: University of Arizona Press, 1962.

WAX, MURRAY L. "The White Man's Burdensome 'Business.' " (Review essay on change and constancy of literature on the American Indians.) *Social Problems,* volume 16, pages 106-113.

WAX, ROSALIE H., and ROBERT K. THOMAS. "American Indians and White People." *Phylon,* volume 22, pages 305-317, 1961. Reprinted in *Native Americans Today: Sociological Perspectives.* HOWARD M. BAHR, et al., editors. New York: Harper & Row, 1972.

YETMAN, NORMAN R., and C. HOY STEELE, editors. *Majority and Minority: The Dynamics of Racial and Ethnic Relations.* Boston: Allyn & Bacon, 1971.

Education for
Constructive Marginality*

Steven F. Arvizu
Sacramento State College

Student of anthropology and education, with extensive experience in teaching and counseling, especially working with young Mexican Americans in California. Director of the Mexican-American Education Project, Sacramento State College, and author of numerous publications dealing with educational research relating to the Chicano movement.

* *Condensed adaptation of a paper presented to a conference entitled "Toward a Philosophy of Education for the Mexican American" at the University of Texas at Austin in 1971.*

To Barbara and the many other Chicanitos to whom schools have been unbearable, in hopes that all of us as teachers can become more understanding and behave accordingly.

I want to examine the anthropological implications in the education of Mexican Americans—indeed in all education—for as we proceed it will become increasingly clear that while anthropology is a tool which has been used to our detriment it can and must be used to our benefit. We shall, in fact, discover that the Mexican American, the Chicano, is a highly advantaged person, for he has the benefits of true constructive marginality, the ability to live in more than one culture, and so to enjoy a richness not possible for those limited in understanding and experience to a single ethnic setting.

Understanding the concept of culture, the process of cultural change, and what happens when two cultures meet enables one to appreciate the impact and dynamics involved in the Mexican-American situation in education. It provides the opportunity to influence constructively the culture of schools, which in themselves reflect the cultural system, and at present reinforce the exclusivity of traditional Anglo-American values as the single valid life pattern in the United States.

Let me tell you something about myself and what I believe. "Let me blow your minds," as those of the counterculture would say—which is an appropriate introduction, since, as you will see, I resent the unimaginative monolingual, monocultural, egocentric, ethnocentric standard of education reflected in traditional instructional program planning. I am a a Chicano educator, teacher-trainer, and student of anthropology. What I present here are my opinions and impressions —an eclectic view of the educational process itself, especially for the Mexican American, as I perceive it. My perceptions and analyses are open to question and examination, of course, as my background of experiences has produced patterns which influence how I behave and how I think, much in the same way as any investigator, educator, or teacher is influenced by his cultural set. For there can be no such thing as true objectivity. We all receive conditioning, we all wear glasses that are tinted by the nature of our particular culture, social group, and lifestyle. Understanding this and recognizing that there is not just one culture, but several, flourishing within this country, it becomes obvious that what is needed is an insight into culture itself, and we turn to anthropology and the study of man and his behavior.

Anthropology is a recent science in the United States, influenced greatly by the classical and traditional sets of the British and Western Europeans. Toward understanding man and how he thinks, speaks, lives, it may be thought of as a tool, a vehicle, an instrument.

> "Es una cosa tener guitarra,
> Y otra saber tocarla."

"It is one thing to have a guitar and another to know how to use it." Or, to put it another way, mastery of a discipline and possession of an instrument do not insure that constructive utilization will result. Anthropology, like any other science, is useful only if it can be applied to contributing to the extension of knowledge or to the solution of problems.

A cruel irony has been the creation and perpetuation of stereotypes by anthropologist and educator alike, who have viewed the ethnic minorities through their own (nearly always Anglo-American) glasses, and found the individual of the Chicano subculture somewhat different (i.e., substandard: of low intellectual ability, having deplorable hygiene, being economically hopeless, with great language difficulty—although, of course, highly talented musically; altogether an inept, if somewhat charming, sort of frito bandito!). Quite a picture—not only unkind but untrue. And morally indefensible. How has this remarkable portrait come about? By applying meaningless devices in evaluating the performance of the Chicano, by insisting on measuring him in white Anglo-Saxon Protestant terms, by refusing to acknowledge the validity—the existence—of his own rich heritage and culture. It seems that the response of educators and social scientists, because of their traditional orientation, has been to concern themselves with the study of the learner's difficulty, as if the student were responsible, rather than to examine the various other factors which influence learning in our schools and to recognize the many elements contributing to the lack of success in education.

I find the present situation in education for the Mexican American and other ethnic minority learners untenable. The use of I.Q. tests on learners from a different cultural, economic, or ethnic group makes about as much sense as would my presenting this paper completely in Spanish. What would that prove? To what purpose (other than to produce a negative self-image in members of minority groups) are repeated studies based on negative observation and inadequate instruments? The tons of research prove nothing other than their uselessness. The reason so many Mexican-American learners do poorly in school is that they are made to feel inferior to learners of WASP background, to feel that they must conform to the dominant cultural pattern, that to do otherwise is wrong—and the penalty turns out to be lower grades, fewer opportunities, and little encouragement to develop as an individual. Simply put, it is cultural destruction. So much for negativism and the misfortune it has brought.

Now let us consider the positive contribution of anthropology toward the goal of education, of providing a beneficial learning environment that will enable the young Chicano to mature into adulthood as a developed human being, along with others in this country. First of all, there is the obvious aspect of his knowing Spanish—indeed, it is often his "first" language. While educators usually look upon bilingualism as a problem to overcome, it should be apparent that it is instead a potential to be developed. Biculturalism is of tremendous advantage—or should be—for it carries with it marginality, the ability to pick and choose constructively and eclectically from two different and useful cultures.

The Chicano is a product of his culture, and he produces his culture. A way of life with all of its patterns exists before a Chicanito is born. These patterns begin to influence him as he develops, which is a natural process that many anthropologists call "enculturation," sociologists call "socialization," and educators call "education." The cultural determinism, the degree to which a person's culture determines his behavior, can diminish, however, with a critical stage of self-awareness, which occurs at different times for different people. At this point, self-determination for the individual takes precedence over group and cultural influences, and the individual gains control over how he chooses to spend his time and energy. He then becomes a producer of culture and can influence for himself and for others what the ever-changing abstract of Chicano culture is going to be.

Anthropology is an obvious tool, a vehicle, by which others may come to recognize, appreciate, and respect the Chicano lifestyle as a legitimate way of life. Listening to Chicanos tell of experiences suffered in education documents realistically how greatly, how profoundly, we need to establish the validity of Chicano culture in widening the mainstream of society in the United States.

Within my own family I can identify with the situation. My language, the food my mother cooked for me, the way my brothers took care of me are very personal and valuable. Can you imagine the dynamics of having these values rejected? How do I answer my daughter when she asks, "Daddy, how come all the kids call me Leah, Leah, Tortilla?"; what do I say when she says, "My teacher doesn't like me; I know, because when I told her I had a drum, she told me 'no, we want an American drum, not one from Mexico.' "? How can I explain my thirteen-year-old niece taking her own life, leaving a note describing school problems as the reason? What about the many other children who cannot talk to anyone about their values and their culture, their language and their ideals? How much do they suffer? What *is* American?

Steven F. Arvizu

If we were to look at a cultural spectrum to identify Chicano culture, we would have to recognize the broad range of patterns that exist between the absolutes (which have never really been defined) of Mexicanness and Angloness. Unlike classical and traditional theories (in regard to marginality), it is proposed for consideration that the Chicano is not "caught between two antagonistic worlds"; rather, he can enjoy the position from which he can function in a variety of language and cultural systems, thereby selectively choosing from various worlds, for it is to his advantage to do so. In this sense he is more free, more in control of how he fits into his environment. This is self-determination at a very personal level and definitely can be viewed as a psychological advantage in today's society.

As teachers become more culturally sensitive and skilled in ethnopedagogy, children will tend to succeed more often, and in a learned, educated way acquire and develop intercultural understanding. Thus it is highly critical that generalizations of the Chicano, especially by educators, accurately allow for the wide range of behavioral patterns which exist within the group.

It might be useful at this point to present one aspect of Chicanoness—language—which to me is distinctive from the dominant culture in the United States. Language, like culture, is a word which carries many meanings to different people. Having similar understandings of meanings of words is crucial in eliminating miscues between one who sends messages and one who receives messages. Often one's background affects the meaning carried by such a word as language. For our purposes here, language means very simply, "a system of communication."

Variations in language (dialects) are normal. Throughout history the way of communicating of a particular area has gained more prestige than that of other areas. The greater the prestige the more desirable a dialect it becomes to learn or use. However, greater prestige is not the same as universal appropriateness. In the case of the Chicano, the "standard Spanish" from Spain, Mexico, New Mexico, etc., has been imposed by many as the power dialect. Anything less than had less prestige and been labeled as substandard ect. Pochismos have thus, mistakenly, been signs of lucated. Educators have missed a most important nding—if two or more people communicate, their system has validity. In the Southwest, millions of e a dialect or dialects of Spanish, and their system of communication is legitimate for their purposes. In fact, public servants who work in barrios are inadequately trained when they study only standard Spanish.

School systems are basically monolingual. Obviously, some Chicanos can function well in such a system. But many are denied equal educational opportunity because of the pattern in our schools which assumes "English is right, everyone speaks English."

Education seems to offer some hope in reorganzing schools and in offering alternative plans for instruction. In fact, bilingual education can offer to non-Spanish speakers that which the Chicano has had for decades—the opportunity to learn the language and culture of another group of people. Respect for difference can be encouraged by giving priority in education to learning the languages and cultures of other groups. According to Sapir, human beings are very much at the mercy of the particular language which has become the medium of expression for their society.[1] Benjamin Lee Whorf presents the theory that language is an influencing factor and shapes the process of perceptual and unconscious thought.[2]

The expression in Spanish, *"Cada cabeza es un mundo"* (each head is a world), tells us that people live in different worlds of reality. In addition, the expression *"Hablar dos idiomas es ser dos personas"* (to have two languages is to be two persons) means that people who speak two languages have two worlds in the sense that the languages they speak affect, to a considerable degree, both their sensory perceptions and their habitual modes of thought.

It is possible to look at structural differences in language to compare ways of life. For the purpose of illustration, we can look at Spanish and English and briefly analyze how they deal with the value of interpersonal relationship to show how linguistic patterns and sensory perceptions, and *"interpre-*

taciones de la realidad," are related.

The first example deals with the matter of courtesy among Chicanos. The Chicano movement tells us to "respect your neighbor, but only let him mess over you once; if he does it a second time, you got to get it on." In looking at the "respect your neighbor" part of the above sentence let us look at Spanish and the concept of courtesy.

In looking at a particular social event—say, introducing yourself to someone—it can be seen that Spanish differs from English in how respect for the other person is treated linguistically. In English, something like "My name is John Doe, I'm pleased to meet you" suffices. In contrast, the Chicano might say, *"Permítame presentarme, Fulano de Tal, a sus órdenes."* There are a number of expressions which denote that the other person is more important when introducing yourself to him, i.e., *a sus órdenes, para servirle, su servidor, a la orden.* To translate these expressions would make them sound stilted in English (at your service, your servant, etc.), yet they are very natural in Spanish. A crosscultural look at introductions (Japanese and other non-Western cultures have an elaborate system) shows that those who speak Spanish are not deviants in this regard.

The expressions *"vamos a comer"* and *"vamos a echar un trago"* denote more than "let's go eat" and "let's go have a drink." In Spanish, the one doing the inviting is assuming responsibility for paying, whereas in English it's Dutch treat, unless it is a courting situation. Let me illustrate this by something that really happened to me in Mexico.

In meeting a young couple at *el teatro,* a group of us were invited by the young man *"a tomar una taza de cafe."* Of the twenty in the group, only a few were native speakers of Spanish. In looking around I noticed that the native speakers of Spanish, myself included, had ordered coffee or tea, which sells for two pesos ($.16 U.S.). All of the others, who were monolinguals from the United States, had ordered Black Russians, Tom Collinses, Scotch and water, etc., which cost about 20 pesos ($1.60 U.S.). It was obvious when the check came that the young man who had invited us was not going to relinquish his unstated obligation to pay. The Anglos had thought that they were going to pay for their own drinks individually or put in a fair amount toward the check, and accordingly had ordered several drinks. They had not understood that the Mexican who had invited us would lose face if he allowed them to pay for our portion of the bill. The concept of *yo invito* is translatable linguistically but not culturally.

Let us take a look at a visit to someone's home. *"Aqui tienes tu casa"* utilizes the familiar, friendly form, making the invitation more personal than "my house is your house." A Mexican family with whom I once lived told me that they didn't understand when a *Norteamericano* used that phrase, for most say it without really meaning it. My friends indicated that in Spanish the expression meant to them "aqui tienes tu casa."

With the Anglo it is understood that you call before you visit, that the wife be allowed to clean up or fix something special for dinner. In some Mexican homes you are invited to sit down and partake of something—water, coffee, food, whatever. In such homes you very seldom leave without having received something from the family. My mother explained that she is extending a part of herself to her guests by offering friendship and hospitality. Some Anglos, when offered something they don't like, simply say, "No, thank you." In the latter situation the appropriate polite thing to do might be to accept even if you don't really want to. Even structurally, in Spanish, the "Thank You" would come first, *"Gracias, no."*

We can gain more insight into how some Spanish-speaking people view courtesy by looking at terms and expressions used by the group. The Spanish language has four different ways of saying you—familiar-singular, familiar-plural, formal-singular, formal-plural. In English one form is used in most situations. As the chief instrument of communication, language attaches words and phrases to the most frequent and most important cultural meanings. We see clearly now how anthropology can contribute substantially toward the improvement of the plight of the Mexican-American student, the young Chicano who is caught—not between two cultures and two languages—but in a web of unfairness that is basic to the present educational system.

Fieldwork training can be useful for teachers in making them less ethnocentric and more sensitive culturally. A program to adequately educate the learner, whose background of experiences are not conducive to learning situations typically offered, must provide for teacher training and in-service which prepares the teacher culturally to meet expected educational challenges successfully. Along with content and discipline mastery, consideration of the language patterns, family structures, economic situation, and other cultural factors evident in the learner's "way of life" should be required of teachers. This is especially true for Chicanos. Chicanos are here to stay, and they are increasing in numbers at a faster rate than most other groups. For many in our country the baby boom is past, and planned parenthood is the byword—not necessarily true for Chicanos. In California, Chicanos are the largest ethnic minority. There are more Chicanos in California than blacks in any other state in the country and we are increasing geometrically. With the eighteen-year-old vote a reality, and the fact that Chicanos as a group are younger than most other Americans, the implications for the future are very clear. Communities are demanding Chicano teachers, or at least culturally sensitive teachers at schools which serve Chicanos. Indeed, it is in our best interest, critical to the common good, that more teachers receive adequate training.

Chicano educators now working as cultural brokers and/or agents of change within the culturally diverse society in the southwestern United States bear our consideration. Educators of Mexican descent who are participating in the dominant culture are key agents to the contact situation. Due to their marginality they are able to identify with both groups. They are the focal point upon which the institution of language and its role in the educational system can be studied. They are looked to as advisors and experts in Mexican-American affairs by the dominant culture, and are expected to stay in the vanguard, protecting the interests of the Mexican-American community. Their influence in both roles, however, is open to speculation.

Education for Constructive Marginality

The perspective of the Mexican-American educator toward language and culture might give insights into the present direction of educational philosophy. Since educational philosophy is basic to instructional programs and their end-products, any element which might affect this philosophy is relevant to the contact situation.

The philosophy of teachers as to how instruction can be improved depends greatly upon how they view the component elements of an instructional program: the learner, the teacher, and the home. The learner is the center of an ideal instructional program, and the test of successful teaching is whether or not his desire to learn is deepened and his capacity and skill to learn are enlarged. It behooves the professionally alert teacher, in order not to be at a disadvantage, to gain information about the learner and the home, and to understand the real world of the child by understanding the lifestyle of the local adult community in which the child lives.

The responsibility of teachers and most public servants today transcends the traditional role. In the case of the teacher, it is proposed that he consider the interdependence of his role as director of learning within the classroom and his role as a community-school liaison person. The role of the teacher should be to take the lead in building an effective system of communication among the student, parents, teachers, and the school. As such he might consider being learner-centered rather than administration-centered. In accepting all children for what they are, the teacher would be an important cog in the wheel for providing an equitable educational opportunity for everyone, preparing all young persons to utilize the benefits of our economic, political, and social systems, and to grow in self-actualization.

Functional teacher-training gives validity to the concept that a significant portion of the training for a teacher should include exposure to out-of-classroom factors and conditions which influence the real world of the child. The educational situation of the low-income student is very much linked to surrounding conditions, and it may sometimes be necessary to change outside class conditions before the learning act can be facilitated *within* the classroom. There is a direct mandate from community groups in the Chicano community that teachers and public servants be aware of social, economic, cultural, and political realities within the Mexican-American communities. The training of the teacher as an innovator and advocate of change as presented here should provide direct results to the benefit of the community in which the teachers are trained and should provide long-term benefits that the traditional role has failed to produce.

The fact that special programs are necessary to satisfy the needs of certain learners implies that the traditional approaches and situations are indeed inadequate. Perhaps some basic philosophical questions as to reassessing the principles upon which educational programs are built are in order. It is entirely possible that by asking vital questions about our schools we may come to realize that such innovations as bilingual education and ethnic studies can effectively take the place of some presently inadequate patterns and plans for instruction while not adding further costs to the educational budget systems.

Several studies [3] demonstrate the relationship of teachers' attitudes and perception of children to the growth and achievement of those children. My mother used to tell me, "Mijo, you see what you want to see." And observation and analysis clearly can be devised to alter the culture of schools themselves for the benefit of culturally diverse learners. The Chicano, Mexican-American, Spanish-speaking, or migrant learner has all of the same basic physiological, social, and emotional needs that other students have. A basic need of man is self-worth, and the minority learner is no different from anyone else in that respect. He needs the educational experiences which will enculturate him without destruction of his cultural heritage, and which will prepare him for adulthood in our changing society. He needs the opportunity to develop computation and communication skills. The self-actualization process will enable him to know himself and to identify with his environment. And his need for economically salable skills need not be neglected, either. The possibilities are great, and we are limited only by our own expectations.

Thus, we have applied the tradition of anthropological inquiry to a contemporary situation for a specific purpose. We have seen how applying the methods of anthropology to education will result in the revitalization not only of the Chicano subculture, but of the larger subculture of which we are all a part—the culture of the schools, of the educational system itself.

The educator has an immense influence on priorities within our society. He is the gatekeeper, trend-setter, and the engineer of what is to come. So much depends upon his priorities.

There is a saying in my family: *"Cuando uno de nosotros sufre, tudos sufrimos."* (When one of us suffers, we all suffer.) When young people take to the streets, to "blow-outs," riots, to "shoot-outs" in East Los Angeles as a means of coping with our social and educational institutions, I ask myself, "What is my responsibility in all of this? What can I do?" What can we as educators do to relate our areas of expertise to resolving this contemporary and obvious dilemma? Truly, we are dealing with a matter of survival.

When we consider the obligation of our educational system—to prepare students to develop their potential to its fullest and to contribute to their society—we must make immediate and long-range plans to educate those whom our system has most failed—most conspicuously, the ethnic minorities. Millions of dollars are spent on the development of natural and technological resources within our society. One of the most valuable of our resources, the human resource, is where more of our efforts should be channeled. The contribution to our country and world will be great indeed when we educate our young for constructive marginality.

NOTES

[1] EDWARD SAPIR, "Conceptual Categories in Primitive Languages," in *Language in Culture and Society*, edited by DELL HYMES (New York: Harper and Row, 1964), page 128.

[2] BENJAMIN LEE WHORF, *Language, Thought, and Reality: Selected Writings of Benjamin Lee Whorf*, edited by JOHN B. CARROLL (Cambridge, Mass.: MIT Press: Moss: 1956).

[3] Among others, ROBERT ROSENTHAL and LENORE JACOBSEN, *Pygmalion in the Classroom: Teacher Expectations and Pupils' Intellectual Development.* (New York: Holt, Rinehart & Winston, Inc., 1968).

Identity

Cultural History and the American Identity
David Brion Davis
Yale University

Pulitzer Prize-winning historian. Specialist in recent directions of American cultural history. Professor of history, Yale University, since 1969. Author of Homicide in American Fiction 1709-1860: A Study in Social Values *(1957),* The Problems of Slavery in Western Culture *(1966).*

A casual contempt for the past has long been a hallmark of American culture. We should not be fooled by the ritualistic homage which Americans pay to historical symbols. In the United States there has seldom been any reverence for the pastness of the past, or for what one British historian has described as "a unity in which the claims of the present are irrelevant." It has been difficult for Americans to believe that the freshest of human experiences have been experienced countless times before, or even that the defining sources of their identity might lie in the past. For over two centuries each generation has considered itself a "now" generation, which was the meaning of Jefferson's famous doctrine "that the earth belongs to the living, and not to the dead."

Although I am speaking of general public consciousness, it is clear that even professional historians have been influenced by the prevailing emphasis on the contemporary, the relevant, the fashionable. This bias has entailed certain losses, such as the use of the past as a flywheel to steady erratic enthusiasms of the moment. But it has also meant that American writers have been less content with the traditional notion of history as a narrative of what happened. Compared, for example, with their British counterparts, American historians have been more open to the methods and concepts of other disciplines. They have also been far more explicitly concerned with searching the past for clues to the meaning or solution of present disorders. They have been happy to assume the role of the nation's psychotherapists, explaining, characteristically, that history is not a chronicle of names and dates, which everyone finds too difficult to remember, but is rather the key to our pyschological identity, which in America is not to be confused with tribal or ethnic identity.

I am aware that I have been using the method of cultural history to account for some of the recent trends of written cultural history. This seeming paradox arises from the fact that historians are partly the products of their culture, and that *their* products in turn become historical subject matter as man continues to reflect introspectively on his own condition and destiny. Thus the American cultural historian has often been a victim of the same present-mindedness and ideological homelessness which he analyzes in a succession of figures from Jefferson to William James. When he argues that our imagination has been historically dominated by such themes as American innocence, the American Adam, or the New World as Virgin Land, he is not calling necessarily for liberation from a belief in liberty or for an abandonment of the Jeffersonian faith that the earth belongs to the living generation. Rather, he is seeking to make us critically self-conscious of historical residues which we take for granted. He is trying to understand historically shaped identities by studying the points of intersection between individual lives and historical processes.

As I have already suggested, cultural history today means a good deal more than the history of taste and style. Yet it remains heavily indebted to the older, pioneering studies of art and religion which identified the characteristic styles, motifs, and patterns of a given period. The achievement of the earlier cultural approaches to history was to point out unities and interrelationships which were not consciously planned and which later could not be verified by any scientific test. Modern scholarship may find an exaggerated synthesis in Henry Adams's medieval France, in Jacob Burckhardt's Renaissance Italy, or in Arthur O. Lovejoy's eighteenth-century neoclassical Europe. Too often such historians restricted the so-called "climate of opinion" to a few major figures, or attempted to use literary sources to explain the functioning of an entire society. Despite these excesses, they succeeded in showing that diverse modes of expression might share common unifying tendencies or organizing principles. And surely one of the fascinations of art history is the way it teaches us to look for stylistic and thematic continuities, not only in painting, sculpture, and architecture, but in poetry, drama, furniture design, and fashions of dress. If we are to search, as I have suggested, for points of intersection between individual lives and historical processes, it is clearly helpful to know how an age saw itself; how it decorated and presented itself; how people perceived and used time and space; how they viewed their own past and destiny.

These questions lead us on to the various anthropological definitions of culture which have had an increasing influence on the writing of history. There is no need here to review the controversies over the culture concept, over its applicability to modern technological societies, or over the discrepancy between culture as actuality and culture as an abstract description of norms and trends. At the risk of gross oversimplification, I will make four brief observations on the application of various definitions of culture to American history. First, there is little ground for an elitist or trickle-down theory of culture, which is not to say that historians should ignore the Jeffersons and Emersons and concern themselves only with collective opinion polls. Second, because of the heterogeneity and the diverse subcultures of American society, one must be on guard against easy generalizations about what Americans believed, feared, or aspired to. When analyzing the American past, we should not expect to find the congruencies and institutional harmonies of thirteenth-centry France or of the timeless Trobriand Islanders. Third, although it may make sense at times to speak of culture as having a life of its own, in the sense of its giving form to experience or creating needs to be fulfilled, there is always a danger of reifying the abstract concept of culture and of endowing it with too much explanatory power. It is necessary to find means of relating cultural patterns to social structure and, more specifically, to such factors as mobility, the density of populations, the division of labor, and the ecology of villages, suburbs, and cities.

Finally, the most significant ventures in American cultural history have moved beyond description and classification, and have stressed the social, linguistic, or psychological functions of various cultural forms. I will later elaborate on the meaning of function, but let me offer a few initial examples. In the nineteenth century many immigrants from Europe were unfamiliar with newspapers. According to Oscar Handlin, the foreign-language press served the social function of helping them adapt to a bewildering new environment while preserving their sense of special identity. Seymour Katz has recently discussed the linguistic function of American fiction, arguing that "works of imagistic literature are ways of knowing in themselves. As such, they also enter into the larger cultural process of forming new concepts, or extending, criticizing or reconstructing already existing concepts." He interprets Hawthorne's *The Scarlet Letter* as a "cognitive model" which gave new meaning to such concepts as adultery, nature, free will, and character. An example of psychological function is the well-known theory that certain nativist and hate-groups sought a scapegoat for the displacement of diffuse aggressions aroused by impersonal social forces.

As seen by recent studies, the history of American culture has not been a succession of harmonious tableaux, each governed by a central motif, but has rather taken the form of dialectical struggle. I do not mean a struggle between metaphysical principles or even between historical systems like feudalism and capitalism. The emphasis, in the most influential studies of the past two decades, has been on conflicts in value and on tensions arising from an interplay between past choices and commitments and new ideas and situations. Often we discover that the incongruity is only apparent or is at least contained within a structure of compatible values. For example, the anthropologist Cora Du Bois speaks of "spurious" conflicts between such American values as conformity and individual success. She suggests that when Americans adopt the posture of friendly, easy-going "good guys," they are attempting to reduce tension generated by the apparent opposition of achievement and conformity. The reason the conflict is spurious, according to Miss Du Bois, is that these are derivative values which are subordinate to America's harmonious commitment to equality, perfectibility, and man's mastery over nature. But systems of values acquire meaning only as the values are acted upon in concrete historical situations. The relationship between the abstract values of equality, perfectibility, and man's mastery over nature depended in part on a complex heritage of symbolic associations, going back to the ancient Greeks and Hebrews, and also on the changing conditions of life in the New World.

With these introductory thoughts in mind, I shall try to summarize and synthesize a few of the major themes of recent writings on American cultural history. From ancient times the idea of nature carried diverse and enormously significant meanings. One of these was a thing's intrinsic character or essential being. Another was the norm, or what a thing ought to be. Still another was the origin or source of a thing's being. These different meanings of nature had long been confused in Western thought, and various kinds of primitivists had looked for universal standards in the simplicity of barbaric or pastoral life. The discovery of a primeval New World could not help but suggest an uncontaminated wellspring of life's true principles. Because the fresh New World was supposedly closer to the moment of creation, or at least to the lost Golden Age, it was a source of regenerative truth.

To these philosophic associations were added the religious meanings of desert, wilderness, and Promised Land. For the ancient Hebrews the desert was at once a refuge from persecution and a place of purgation and consecration. It was in the desert that the covenant was sealed and that Israel became dedicated to her mission. And it was in the midst of the desert that a new paradise was to bloom. In Christian symbolism both the monastery and church were likened to a garden in the wilderness. This rich heritage of religious symbols gave prophetic meaning to the discovery and settlement of a virgin continent. From the time of the Puritans to the time of the Mormon pioneers and beyond, the American wilderness was a refuge, a place of trial, and a garden where the invisible church of saints would one day become visible. In short, it was the stage where the last act of man's salvation would be played out.

There was, however, a fundamental antinomy in the ideas of nature and perfectibility which the Old World projected onto the New. The path of historical progress seemed to lead to a primitivistic regression; or, to put it the other way around, the moral regeneration which was to come from purification and simplification was to be linked with the conquest and settlement of a continent. How was Western man to be morally restored by nature without destroying the source of his restoration? How was he to avoid lapsing into unenlightened savagery unless he rebuilt on American shores the very institutions which he associated with sin and imperfection?

One of the first historical studies to deal with these questions on a grand scale was Henry Nash Smith's *Virgin Land* (1950). Whereas earlier scholars had focused considerable attention on the actual settlement of the West and had endlessly debated the importance of the frontier in shaping the American character, Smith was concerned with how Americans had perceived the West, or, to use his words, with the West as symbol and myth. Although his use of these terms remained imprecise, he succeeded in showing that imaginative responses to the West have left a deep imprint on American culture. For our purposes, his most important finding was that the symbolic West reflected a vast network of conflicting expectations. Thus, the West was long seen as the "passage to India" and the route to imperial expansion and power. The West, in short, was to be the key to America's greatness and would allow her ultimately to dominate or "save" the world. But the same region was cherished as a refuge, sanctified by nature, from the corruptions of civilization. Similarly, the heartland of the continent was described as a bountiful garden that would supply the entire world with food and wealth. As a source of unlimited abundance, the West promised to rid mankind of poverty and want. Yet it was pictured as a self-contained retreat that protected the simple virtues of agrarian democracy and provided a safety valve for the dangerous pressures of the urbanizing East. These conflicting values became personified in the figure of the western hero, as he evolved from Daniel Boone and Leatherstocking to the modern cowboy. On the one hand, the primeval hunter, scout, or mountain man leads the vanguard of civilization. He helps subdue the wilderness and prepares the way for the wagon trains, the land speculators, and the farmers. Even the cowboy liquidates the Indians and outlaws; he is the champion of law and order. But on the other hand, the western hero is a child of nature, the American Adam, who is in retreat from the sordid complexities of organized society. His virtues derive from his simplicity and from his communion with primitive nature. Smith shows that western mythology has had a continuing influence on public policy. His analysis deepens our understanding of American isolationism and its curious links with Pacific expansionism. We come to appreciate some of the psychological ties between western symbolism and the American male ego (though advertisers of cars and cigarettes have not had to learn anything from Henry Nash Smith). We are still unenlightened, however, as to the place and function of such mythology in the larger culture. It is also unclear whether the mythological West was a unified construct which served vital social functions precisely because it held contradictory desires in delicate balance, or whether the imagined West was a fragmented mosaic of images which reflected a confused and disordered culture.

Two studies which extend Smith's thematic polarities are R. W. B. Lewis's *The American Adam: Innocence, Tragedy and Tradition in the Nineteenth Century* (1955) and Charles L. Sanford's *The Quest for Paradise* (1961). Although Lewis was concerned with the literary and philosophic "dialogue" over the meaning of America's newness, he was willing to take these "clashes over ideas" on their own terms, without resorting to social or psychological explanations. His main point takes us back to the antihistorical attitude which I discussed at the beginning of this paper. It was precisely because the New World was seen as promising a new beginning for mankind that many Americans considered themselves emancipated from history. The collective myth of the

American as a new Adam, a man of innocence who is free from the limitations of past decisions, forced the literary mind to search for new symbols that could help account for the actual complexities of American life. American writers could not rely on the traditional scale of values with which European literature had interpreted the human experience. In response to the "Party of Hope," which upheld the American faith in innocence and perfectibility, there emerged a "Party of Memory," which looked nostalgically to the lost romance, heroism, and grandeur of the Old World. As might be expected, this dualism became resolved in a "Party of Irony" whose creative energies depended on both a sympathy for the American Adam's naive aspirations and a knowledge that his fall is inevitable. Presumably, the Party of Irony would also include those cultural historians who believe in American ideals while demythologizing their historical foundations.

Sanford, for example, interpreted the supposedly unique American mission as a secularization of the traditional Christian mythology of redemption, in which the West came to stand symbolically for a material paradise, and the East for hell. He also suggested that the American impulses toward expansion and escape, toward millennial mission and primitivistic withdrawal, can ultimately be explained by more universal biological impulses which lead man to self-assertion, followed by regression toward an infantile Nirvana.

Whatever the biological origins, it is evident that Smith and Sanford are dealing with two basic varieties of human experience. The first includes outward expansion, a confidence in the effectiveness and righteousness of one's power, and a belief in the progressive nature of change. The second can be described as an inclination to withdraw and turn inward, a fear of being oppressed, corrupted, or contaminated, and a desire for self-purification. In Europe, for the most part, there were institutional limits to both varieties of experience. But in the New World there were few boundaries for either *conquistadores* or perfectionist sects.

It is important to remember that America's antihistorical consciousness originated in the ideals of individual freedom and moral autonomy of the European radical sectarians. I have in mind a variety of dissident religious groups, including the Puritans and Quakers, who drew on a long heritage of struggle against ecclesiastical establishments. There was, it should be stressed, no problem of identity for the European sectarian. He was the voluntary member of a group that had ultimately been chosen by God to restore the lost purity of the Christian Church. If he professed faith in the liberty of conscience, in the moral autonomy of the individual, and in the inevitability of the millennium after a period of struggle with the powers of darkness, he also stood in the position of a heretic condemned and persecuted by the established order. The sectarian ideals acquired social meaning and served as forces for unity and cohesion when they were pitted against the opposing power of a vast institution or corporate body. And the sectarian's mood of expansionism and mission alternated with the need for quiet withdrawal and self-insulation.

In America, however, the way was open for expansion, change, and self-determination. Paradoxically, this removal of restraints had the effect of combining self-assertion with withdrawal and self-purification. This was, in effect, one of the dilemmas of the American Puritans, who tried for a time to maintain the fiction that their temporary "errand into the wilderness" was the most effective way to transform English society and religion. Yet their very success as builders and self-governors cut them off from the society they intended to reform. To turn inland, in order to explore and settle the continent, necessarily meant turning inward, in the sense of turning one's back on the past seats of civilization. And, of course, the impulses toward expansionism and withdrawal came to a head with the War of Independence, which was the supreme act of dissociation and severance, but was also a portentous sign that America's political power would promote and protect the expansive energies of the economy. The two tendencies merged repeatedly in the thought of a man like Jefferson, whose fear of contamination from Europe was matched by his confidence in America's westward-turning destiny.

By the time of the War of Independence, the sectarian

ideals had been gradually appropriated and secularized by the international Enlightenment. Both in their religious and political versions, they formed the basis of a vague American ideology. To uphold the values of equality, perfectibility, and man's mastery over nature was thus to uphold an ideological establishment. Yet by their very nature, these values of individualism were diffuse and anarchic. They could lead, if pushed to the extreme, to a repudiation of all laws, mores, and social norms that limited the self-affirmation of the individual. As both the Puritans and Founding Fathers well understood, it was necessary to find new American equivalents for the institutional restrictions and restraints of Europe. And as Alexis de Tocqueville later sensed, this was the American people's greatest burden, as well as achievement.

Moreover, the uprooted Puritans and Quakers soon discovered that while there were no kings or bishops to fear in America, there were also no effective ways to combat an erosion of zeal, a dilution of religious identity, and a growing absorption with material achievement. For each pioneering generation, the underside of Americanization was a losing struggle against their children's loss of faith, exogamous marriages, and movement to greener pastures. Each immigrant group's mission of liberation became fused with the general American quest for making good and keeping up with the Joneses. In contrast to Europe, the idea of mission was not linked with ethnic reunification, with the desire for revenge, or with the reclaiming of sacred ground. The sense of being American—of being part of the American experiment—was far more catholic and open-ended. It arose from a sharing of common experiences.

Historians have generally taken a very positive view of this creation of identity through shared experience. It has often been said that the strength of American democracy depends less on a noble political heritage than on the fact that Americans are made, not born. And it has been reassuring to think that virtually anyone can learn to be an American, from Carl Schurz and Andrew Carnegie to Wernher Von Braun.

What needs to be added, however, is that the mask of American identity can be as coercive as it is superficial. In accepting someone as American, we tend to rely on the lowest common denominators, or rather on those traits of character and behavior that are easily learned in public school and from the mass media. I know a taxi driver in Rome who could pass as an American anywhere, not only because of his speech and his knowledge of National Football League statistics, but because of his un-European posture, gestures, and "good-joe" friendliness. Yet he had lived only three or four years in the United States. If he had refused to be so adaptable and had also continued to live in America, he would have felt the full coerciveness of our supposedly permissive culture.

By coerciveness I mean something more than pressures for conformity. By linking American identity to shared experiences, we have created pressures for expropriating the particular experiences and heritages of many groups. My Roman taxi driver might feel an ethnic pride in the fact that Columbus was an Italian; yet Columbus belongs to us all. Successive generations of schoolchildren, including Indians, blacks, and immigrants, have been exhorted to identify themselves with the New England Pilgrims. Abraham Lincoln is everyone's hero—but so is Robert E. Lee. It is difficult to imagine the English similarly canonizing Bonnie Prince Charlie or Charles Stewart Parnell. Yet soon after our own Civil War, both the Confederate flag and "Dixie" became part of our proud national heritage. In America, it would seem, everyone ultimately wins. Rebels and dissenters are quickly absorbed and neutralized. Sitting Bull and George Armstrong Custer were both good guys who contributed, in some vague way, to a better America.

This sanguine view of the past, standardized in countless schoolbooks, implies that historical conflicts are devoid of moral meaning for the present generation. As Oscar Handlin has observed, Robert E. Lee was personally faced with a profound issue of loyalty, since he was a devoted Virginian and also a West Point graduate and general in the United States Army. In retrospect, however, Lee's dilemma was soon obscured by romantic mythology. Whichever choice he made, he could not really lose. Regardless of who won or who suffered, everything worked out for the best. Much as the competition of consumer products has been said to increase the market for all, so the squabbles of diverse American groups have supposedly contributed to a more prosperous tomorrow. Each scream of protest has been absorbed into the Great American Heritage. We thus have little patience with groups that harbor historical grudges. Nat Turner must be culturally nationalized along with Geronimo and Big Bill Haywood.

It may appear that I have digressed rather far from my initial thoughts on antihistorical consciousness and the "fundamental antinomy in the ideas of nature and perfectibility which the Old World projected onto the New." My aim, however, has been to indicate some of the troublesome consequences of early American attitudes toward history, mission, and national identity. The myth of the American as a New Adam, stripped of all the encumbrances of the past, has made it difficult to perceive that present grievances may be the fruit of significantly different historical experiences. This blindness has been aggravated, moreover, by our voracious desire to absorb and assimilate all cultural diversity. As I have tried to suggest, this cultural imperialism has been related to a quest for commonality in the face of disintegrating ethnic and religious identities. Our sense of national integration, which is surely a good thing in itself, has been bought at the price of a mythologized past that obscures certain realities of conflict and exploitation.

Recent studies in American cultural history have pointed out other ways that collective myths have disguised conflicts in value. This is not to say that American myths and symbols have served no positive functions, although it must be confessed that today's historians have shown little interest in the values of maintenance and equilibrium. But whether for good or ill, the leading American myths or "persuasions," to use Marvin Meyers's term, have often been propagated as antidotes to acquisitive materialism. It is true that Americans have commonly extolled material achievement as a liberating and leveling force. A man who proved his character in the hard competitive struggle, a man like Henry James's archetypal American, Christopher Newman, had no need to feel ashamed of his lack of family, education, or social connections. On the other hand, when material achievement is unqualified by higher goals or by an ultimate transcendence of wealth, it has suggested a pursuit of pure self-interest in defiance of the common good.

Accordingly, Americans have frequently felt the need to legitimate individual self-interest by associating it with a higher theme, such as the restoration of a natural order of agrarian simplicity. Thus Marvin Meyers, in his brilliant study, *The Jacksonian Persuasion* (1957), argues that the followers of Andrew Jackson were really aspiring capitalists, committed to material progress through the widening of individual opportunity. Yet they styled themselves as the restorers of a simple, republican tradition. Meyers calls them "venturesome conservatives," since they were always afraid of losing what they had gained, and also sought to mitigate their acquisitive impulses by appealing for unity against institutional forces that supposedly threatened a noble heritage of republican ideals.

William R. Taylor's *Cavalier and Yankee* (1961) is similarly concerned with collective symbols that reconcile conflicting values. But Taylor is less interested in contradictory attitudes toward past and present than in the creation of character models that transcend individual self-interest. His thesis, if we may oversimplify an immensely complex argument, is that Americans of the antebellum era faced peculiar problems of self-definition. Almost universally, they congratulated themselves on their newly won freedom, their amazing prosperity, and their prospects for infinite growth and happiness. Yet they felt cut off from the rest of the world—not only from the despotic governments and social corruptions of Europe, but from the forms and standards that had given meaning to life. More specifically, they were perturbed by questions of class and leadership. Who was to take the place of the well-born and cultivated classes of Europe? Without a gentry, where was one to look for standards of gentility? Without nobles, how was one to know the meaning of nobility? The American Revolution had endowed the nation's first generation of leaders with an aura of selfless grandeur. But the French Revolution and Napoleonic wars raised new questions about the virtue of self-made political leaders and the wisdom of the popular will. In the beginning of his book, Taylor makes much of the dialogue between Jefferson and John Adams, in the last years of their lives, over the American need for identifying a natural aristocracy.

At least among certain strata of the American population, there was a growing fear by the 1820s that the highest human qualities would soon be submerged in an acquisitive, materialistic civilization. A tendency toward national introspection, toward a self-conscious pondering of the national character, was reinforced by the sneers of European travelers in America and by the discomfort of Americans, particularly men-of-letters, who traveled abroad in search of culture. The stereotyped Yankee, it soon became plain, was a money-grubbing trickster whose mind never rose above schemes for profit. The stereotyped southern planter was not only the coarse descendant of vagabonds and indentured servants, but he now led a life of indolent dissipation, supported by the blood and sweat of Negro slaves.

According to Taylor, the myths of the southern Cavalier and "transcendent Yankee" were "defensive fictions" designed to counter these stereotypes and the self-doubts they engendered. The southern Cavalier, who was largely invented by northern writers, was one of nature's noblemen. Self-disciplined, living close to the soil, he was aloof from the competitive struggle. A compassionate master, he combined the virtues of feudal lord and American frontiersman; yet for this very reason he was out of step with the urbanizing and industrializing march of history. Taylor observes that even the northerner could vicariously identify himself with the southern plantation legend and thus assure himself that he belonged to something more than a purely acquisitive society. But the stigma of selfish materialism was also veiled by the figure of the transcendent Yankee, "a combination of Puritan and Statesman," in Taylor's words, who sacrificed personal ambition for high-minded patriotism. Although Taylor looks primarily to the "free fantasy" of literature for the meaning of these myths, he suggests that such fiction was a projection of social needs and that literary myths had a feedback effect as statesmen like Robert W. Hayne and Daniel Webster played the roles of Cavalier and Yankee. Ironically, these character types, which originated in a desire for disinterested nobility, gave substance to the idea of a divided culture and may thus have contributed to sectional pride and jealousy, and ultimately to the Civil War.

There are interesting parallels between Taylor's study of ideal character types and Leo Marx's *The Machine in the Garden* (1964), although the latter is concerned with America's imaginative response to early industrialism. Like acquisitive self-interest, industrialism threatened an entire scale of traditional values, and particularly the inclination to associate vice with artificiality, and virtue with pristine nature. Marx's central question, then, is how mythology enabled Americans to reconcile technological progress with their characteristic belief in nature as the source of universal truth.

He finds the key to this question in the ancient European literacy convention of pastoralism. By pastoralism he means an "imaginary landscape of reconciliation" between primitive nature and urban civilization. The tradition of European pastoralism had included not only an idealization of bucolic serenity but a larger literary design which, by acknowledging the threatening force of either urban power or untamed nature, introduced elements of irony and ambivalence. In Europe, this symbolic means of harmonizing values had been largely confined to the world of letters. In America, however, the pastoral design acquired sociological relevance from the continuing process of settlement and from the tangible existence of wilderness, on the one hand, and urban industry, on the other. Thus, what once had been a literary device for reconciling abstract values became, in the New World, an ideological instrument for interpreting the meaning of rapid social and economic change.

According to Marx, the middle ground of pastoralism provided a setting for the assimilation of technology to nature (one thinks, for example, of the Currier and Ives prints of Mississippi steamboats and of trains crossing meadows, in which the harshness of iron and smoke is softened by peaceful mists and romantic skies). Confronted by an event unique in history—the penetration of the rural landscape by the machine—Americans avoided a sense of conflict by appealing to the pastoral metaphors of reconciliation. Popular leaders like Daniel Webster, the transcendent Yankee, used a rhetoric which Marx labels "the technological sublime" to make industry merge with nature. And because technological progress could thus be imaginatively dissociated from urban Europe and adjusted to the middle landscape of American democratic pastoralism, Americans could picture themselves as pursuing rural felicity while being dedicated to wealth, power, and production; they could think of themselves as immune from the force of history while leading the world in historical progress.

Both Taylor and Marx are aware of the danger of confusing literary values with the attitudes of various elites or the population at large. They attempt to show, and I think convincingly, that literature is an intrinsic part of life, no less "real" than the rhetoric of politicians or the ideas we associate with the American West. They also give substance to the theory that a single work of art may give symbolic expression to the hidden trends and contradictions of our culture. No doubt they are right in suggesting the remarkable hold which myth has had on the American mind. Yet there is a danger, I would submit, in mythologizing myth. We must guard against placing too much emphasis on what Karl Mannheim has termed "false consciousness." The very term "myth" is unfortunately vague, since it implies falseness as well as simple legend, rationalization as well as creative function.

As a way of concluding this frankly speculative paper, I should like to point to two theoretical strategies which might help clarify the meaning of myth as well as refine the cultural approach to history. The first strategy has to do with the way cultural symbols and metaphors have been assimilated and used by specific individuals. This knowledge could be reached by collective biography or the study of a group of individuals who are in some sense representative of a class, an occupation, an interest, or a mode of protest. Obviously, this kind of study must gain some control over variables by determining the place of the representative men within the social structure. Another approach, demonstrated by Erik Erikson's *Young Man Luther* (1958), a book enormously suggestive for the theory of cultural history, is to examine in detail how the personality crises of a complex individual reflect tensions within the general culture and how the individual's resolutions of conflicts within himself lead ultimately to transformations within the culture. Clearly, the world has known few Luthers, and biography always runs the risk of exaggerating the historical importance of individuals. Biography may provide, nevertheless, a concreteness and sense of temporal development that most studies of culture lack. And by showing how cultural tensions and contradictions may be internalized, struggled with, and resolved within actual individuals, it offers a promising key to the synthesis of culture and history.

The second theoretical strategy concerns the hazy concept of secularization, or what might more broadly be termed cultural transference. While it has long been recognized that nationalism, for example, bears many of the qualities of a religion, it is hardly sufficient to say that nationalism is simply a secularized religion. Despite all that has been written on the subject by sociologists, historians lack a theoretical framework for explaining how new ideologies acquire the functions and some of the characteristics of older cultural forms. Many of the themes and myths that I have discussed here involve a process of secularization or cultural transference. Thus, I casually stated that the ideals of certain sectarian religious groups were secularized by the Enlightenment. Yet what does secularization really mean, even if it could be demonstrated, as I think it can, that there was a historical continuity of ideas? Clearly, the dissenting religious groups of the sixteenth and seventeenth centuries would have been horrified by the later secular reincarnation of their ideals of freedom and perfectibility.

Although analogies are always dangerous, and particularly analogies between social and individual life, it may be helpful to imagine a mildly neurotic man who has acquired a set of so-called "symptoms" that enable him to cope with the world. I am not thinking of the symptoms as evidence of any disease, but only as a systematic, though highly idiosyncratic, way of ordering experience. Now suppose that some new situation arises which places an untenable strain on our neurotic's accommodation with reality. His symptoms no longer "work." As an interrelated system, they are now quite ineffective. Without assuming that our neurotic is becoming either "healthier" or "sicker," we could predict that he will have to reformulate his mode of coping with the world. His new syndrome may include some of his old symptoms that now perform new functions; he may well have acquired new symptoms that carry only faint imprints of the former ones. The main point is that the system itself will have changed.

I think that some model of this sort may help us understand cultural transference, in the purely structural sense of form and function. If I am right, historians might well focus more attention on the transmission and adaptation of ideologies that shape, repress, or redirect individual aspirations. I have in mind what might be termed the "cultural superego," as defined and transmitted by the governing groups of society. This is not a metaphysical abstraction, but is, rather, the orientation conveyed by parents, teachers, clergymen, laws, and social rituals. The crucial question, at each point in time, is whether acculturation (or indoctrination) is relatively complete, or whether new strains, both external and internal, give rising generations the chance to reformulate the values that give meaning to life.

Modern Identity: Crisis and Continuity

Peter L. Berger

Rutgers University

Sociologist of religion, humanist, and Lutheran layman. Extensive teaching in the United States and Germany. Professor, Rutgers University Graduate School, since 1970. Past president of the Society for the Scientific Study of Religion and resident member of the Council on Foreign Relations since 1971. Author of The Noise of Solemn Assemblies *(1961),* The Precarious Vision *(1961), Introduction to Sociology (1963), The Social Construction of Reality (with Thomas Luckmann, 1966), The Sacred Canopy—Elements of a Sociological Religion (1967), A Rumor of Angels—Modern Society and the Rediscovery of the Supernatural (1969), The Homeless Mind—Modernization and Consciousness (with Brigitte Berger and Hansfried Kellner, 1973).*

Recently, moved by the desire to escape the *New York Times,* I started to reread Thucydides' *Great War.* The escapist motive, as I should have known, was emphatically frustrated. Instead, I was plunged right back into just the sort of "relevant" material that I had wanted to get away from— down to Pericles expounding the domino theory to the Athenian assembly prior to its declaration of war against Sparta and, after the desecration of the Hermes statues on the eve of the expedition to Sicily, the great patriotic purge that only by an oversight failed to be called the "Honor Athens" campaign. In the light of these disagreeable relevancies I returned, with more discomfort than on previous occasions, to the famous passage in Thucydides' own introduction to his book. I quote in Rex Warner's translation:

It will be enough for me . . . if these words of mine are judged useful by those who want to understand clearly the events which happened in the past and which (human nature being what it is) will, at some time or other and in much the same ways, be repeated in the future.

Why the discomfort? In the event, to be sure, because of my frustrated escapism. There is also, however, a theoretical discomfort. It is the discomfort felt upon encountering a viewpoint that flies directly in the face of one of the most cherished notions of modernity—that of the ever-changing, ever-innovating character of historical reality—a notion conceived in the secularization of biblical eschatology, born in the revolutionary turmoil of the modern era, theoretically baptized by Hegel, and today part and parcel of the cognitive instrumentarium of almost everyone from Barry Goldwater to the New Left intelligentsia. I like to think that my own *Weltanschauung* is something less than completely modern, yet my modernity reveals itself in the discomfort I, too, feel upon being told that, when all is said and done, nothing in history is really new. On top of that, I am, as it were, professionally obligated to be uncomfortable with such a notion. Let me repeat the key phrase in the passage from Thucydides: *"Human nature being what it is."* A moment's reflection about this statement is liable to make any sociologist acutely nervous, as it seems to threaten the very foundations of his discipline.

Historians, of course, feel differently about this. Except for those contaminated by too much interdisciplinary contact with the social sciences, historians generally react with positive glee to any suggestion that, say, contemporary America is just like Periclean Athens—plus a couple of minor addenda, such as helicopters and television. Sociologists, by contrast, have a deep vested interest in the minor addenda. Their entire professional ideology leads them to the position that the addenda of modernity constitute startling *nova*, profound transformations in the very texture of human existence and human consciousness. Spiritually, almost all sociologists are Hegelians, in that they tend to look upon human nature either as a myth or as itself the product of socio-historical processes. This basic discrepancy in viewpoint between historians and sociologists is likely to come to the fore very quickly in any discussion of the present topic. The sociologist is likely to take the so-called identity crisis of our time with deadly seriousness and to seek explanations in terms of this or that alleged *novum* of modernity (television, if not helicopters, being a case in point). The historian, on the other hand, is apt to fish out some ancient text (*how* ancient, of course, will depend on his sub-specialty), which is supposed to demonstrate conclusively that all of this has happened before in very much the same way.

Who is right? Thucydides or Hegel? Sociologists are also known for a tendency to make rash judgments—I would have to be not only rash but downright deranged were I to suggest that I can resolve this question here (or, for that matter, elsewhere). A good case can be made for the statement that modern thought has discovered history (and thus society) *as against* nature—a discovery contained in Giovanni Battista Vico's classic formulation of the difference between history and nature (we have made the former, but not the latter). But it would be rash indeed to maintain that modern thought has also discovered just where the one ends and the other begins. The only sane attitude in these matters is one of great caution. I would like to approach the present topic in such an attitude of caution, thus disappointing from the beginning all those who expect the sociologist to engage in fiery culture-prophecy. There can be no doubt that what is currently called the crisis of modern identity is a real phenomenon—minimally real in the sense of W. I. Thomas, that anything defined by people as real *is* real in its consequences. All I can do, then, is to look at this phenomenon in the perspective of sociology and to reflect, however tentatively, as to which of its elements are genuinely new and which are in continuity with the past.

Permit me to begin on a fairly abstract theoretical level: What, in the perspective of sociology, is identity as a phenomenon? I believe I am correct in thinking that the current vogue of the concept of identity, and of various theories about its alleged permutations, was begun by Erik Erikson. This is not the place for a discussion of Erikson's highly intriguing work, but it should be emphasized that Erikson's theoretical frame of reference comes from psychoanalysis rather than from any social science. I don't think that this frame of reference can simply be taken over by the sociologist. Minimally, it will have to be considerably modified in order to be useful for purposes of sociological analysis; maximally (which is my preference), the sociologist will try to generate his own frame of reference for the phenomena in question, an enterprise for which the conceptual tools are available in a tradition of sociologically oriented psychology derived from George Herbert Mead. In what follows I'll try to sketch this frame of reference in very broad outline (inevitably, this will entail coming out with axiomatic propositions that cannot be validated in the present context), and then to look at the current situation in this perspective. The perspectival aspect of this procedure ought to be stressed—in other words, it should be very clear that the sociologist is in no position to produce a metaphysic that will answer the philosophical questions of the ages; all he can do is to draw out the implications of his peculiar and necessarily limited insights into human reality.

Identity is a notion obviously related to that of the self. As a concept, it seeks to circumvent the ontological problems suggested by the latter term, yet it clearly has something to do with the question "who am I?" Sociology is an empirical science. As such, it cannot deal with this question in its ultimate philosophical or even religious dimensions; it can only deal with it insofar as it refers to phenomena available to empirical methods of inquiry. Empirically, there are two such referents—the objective social structure within which the questioner exists, and his own subjective consciousness. In other words, sociology cannot deal with the question "who am I *really*?"—as before the face of God, or in the realm of things-in-themselves, or in any realm that is sheltered from the vicissitudes of collective or individual definitions of reality. Consequently, a sociological approach to these matters will have to distinguish between objective and subjective identity, between who I am *to others* and who I am *to myself*. Both aspects of identity are empirically available; neither refers to the self as a metaphysical entity.

Society *assigns* identities. That is, society tells me who I am. This process of identity assignment begins at birth and continues through life. It is an essential element of all so-called socialization processes, not only in infancy and childhood, but in adult life as well. It may be graphically represented as society pinning identification tags on everyone who is part of the societal *dramatis personae*—from then on, or until further notice, everyone will be systematically treated as whatever the tag says he or she is. What is more, everyone will also be expected to take this assigned identity seriously, to play its appropriate role with inner conviction, or at the very least to pretend to do so. It should be stressed that this process of identification is essential to any human society. Without it, we literally wouldn't know from one moment to the next who it is that we're dealing with in our everyday lives with others, and society as an ongoing experience would be impossible.

This may sound rather mechanical, as if the individual were stamped out of a mint. Anyone who has ever dealt with small children knows that this is not how the process works. While Margaret Mead is regrettably correct in her description of socialization as a game in which, almost always, the adults win in the end, the child is not a passive victim in all of this. While the adults are indeed the stronger party, they are far from omnipotent, and the game ends not so much in a victory as in the successful conclusion of a series of compromises. Thus, socialization may be viewed as a bargaining process between society and the individual (a fancier term for this is "dialectic"). Society *assigns* identities —individuals *appropriate* identities. Most of the time, indeed, they appropriate, with greater or lesser modifications, precisely the identities that have been assigned to them— in other words, they inwardly identify with their tags—but there is no complete certainty about this and there are enough cases of noncooperation to make things interesting.

The distinction between objective and subjective identity may now have become clearer. It is probably also clear that the process is much more complicated than sketched here and than we can possibly discuss in the present context. But let us go on. Implied in the above is the possibility of conflict between objective and subjective identities. Society tags me as female, but I experience myself as male, or would like to. Society assigns me the identity of organization man, but I *know* that I'm an artist. Society calls me a nigger, but I assert myself as a black man. And so on. In such conflicts, of course, there is an inner as well as an external contestation: if I am to resist a certain tag, I must not only fight all those others who want to keep it pinned to my chest, but I must also fight myself—or, more accurately, that part of myself that still believes in the tag. How much conflict of this kind there is in a society will depend on historical circumstances that cannot be laid down *a priori*. There will always be some conflict, if only because socialization cannot completely or conclusively mold the biological constitution of the individual. In other words, if nothing else, the probability of "maladjusted" individuals is biologically given. At the same time, any society that functions fairly efficiently will have a high degree of symmetry between objective and subjective identities; that is, *most* individuals will have inwardly identified with *most* of their tags. Put differently, most actors in the societal drama will, more or less, perform with conviction. Since, even in the best of theaters, the management can never be completely sure of this, there are always certain standard operating procedures for those actors who forget their lines, who ad-lib willfully, or who go berserk and threaten to explode the whole show. These procedures are what sociologists call "social control," the details of which do not concern us here.

Subjective identities differ in stability. There are many individual reasons for this, from physiological peculiarities to the biographical accidents that psychoanalysts are interested in. But there are also social conditions for stability or instability. Once one has grasped the dialectical process in which identity is generated, one can say *a priori* that the stability of subjective identity is determined by the stability of objective identity assignment. Put more simply, I will be certain of who I am to the extent that society is consistent in its treatment of me. Such consistency, in turn, depends upon the cohesiveness and continuity of the institutions within which the identity assignments take place. For example, a class of people that is secure in its position in society—and that has been secure for a long time—is likely to produce individuals who are calmly certain of who they are. Such is the proverbial case with aristocrats, but very much the same inner certainty regarding identity can be observed in primitive or peasant societies, or among the members of highly cohesive, closely knit minority groups.

Such certainty should not be confused with happiness. I may know exactly who I am, and be quite unhappy about it; but in that case, unhappy or not, I will at any rate be spared the particular miseries of what we call identity crises. Conversely, unstable institutional contexts are likely to produce individuals who are uncertainly nervous about just who they are. For example, classes of people that are undergoing either a sharp rise or a sharp decline in social position are unlikely to be blessed with stable subjective identities. Naturally, this instability will manifest itself in social action as well as in consciousness.

An episode from the nineteenth century illustrates this point economically. On the occasion of a state visit to England, the Empress Eugénie (upstart royalty if ever there was one) went to the opera with Queen Victoria. Both women cut rather magnificent figures. Majestically, and with perfect aplomb, they went up to the royal box. Eugénie came in first, graciously bowed, graciously looked behind her, and sat down on the chair that a lackey was pushing under her posterior. Then Victoria followed, graciously bowed, and sat down. Victoria did not look behind her. She knew that the chair would be there.

Identity is grounded in socialization. That socialization takes place within an institutional context, which has a particular history. Thus, identity is finally grounded in history, *has* a history or, if one prefers, is a historical product. Psychoanalysis has made us aware of the importance of biography for identity. But every individual biography only makes sense as part of a larger chronology, which is the history of a particular society (and, incidentally, it has been the great merit of Erikson's work that he has elaborated, within his own theoretical frame of reference, which is quite different from the one presented here, the interconnections between biography and history). For example, it was the children of the bourgeoisie who brought forth the luxurious psychic complications that so intrigued Freud—and, at least in principle, he was probably correct in ascribing this or that complication to this or that Herr Papa (not to mention this or that Frau Mama). But *all* these individuals shared a common bourgeois context, were the inheritors of a long history that lay back of that context, indeed (whatever their individual eccentricities in the administration of the incest drama) participated in a common bourgeois identity. If you like, *that* was the large box out of which all the individual identity tags were pulled. But this bourgeois identity did not fall from heaven; neither was it contained in the nether depths of the unconscious from the start of time. Rather, bourgeois identity emerged as the product of the very specific history of the bourgeoisie in modern Europe. Norbert Elias, in his monumental work *Der Prozess der Zivilisation*, has traced this history to its roots in the minutiae of manners and morals. It was a long and painful history, during which, for example, individuals were first taught to stop spitting on other people's plates, then not to spit into their own, then to refrain from any public spitting—a process of territoristic socialization also extending to the regimentation of eating, defecation, sexuality, and speech—finally placing on the stage of history individuals sufficiently "repressed" to be capable of Freudian neuroses in the first place.

It may seem that our theoretical considerations, so far, have led us to a position diametrically opposed to Thucydides'. What place could possibly be occupied by a "human nature" in this scheme of identity productions? Frankly, I don't know; in any case, I am not able to present a philosophical anthropology that will, among other ingredients, contain the aforementioned sociological angles. However, the anti-Thucydidean character of the aforementioned perspective is modified as soon as we are willing to concede (as, I think, we must) that the variety of ways in which identity has been historically defined and produced, vast though they are, yet are bound by parameters. In other words, man does indeed (as Marx put it, in his version of Hegelianism) produce himself—but there are limits to the range of possible products, and these limits may be viewed as transhistorical and cross-societal constants. Some of these constants are given in man's biological (and, possibly, psychic) constitution. I don't think that either human biology or psychology has reached the stage where it can tell us what these constants are with any degree of assurance, though the work now being done by Konrad Lorenz and his school in comparative ethology, for instance, and by Noam Chomsky and others in structural linguistics, is highly suggestive of the directions in which answers to this problem might lie. However, even at this stage of our knowledge about man's constitution, when we do not as yet know very much about the instinctual parameters of his behavior or about the putatively necessary structures of his consciousness, it is clear that this constitution sets limits to the possibilities of man's reality and self-construction. Take only such basic constitutional factors as man's mortality, his sexuality, or his very limited capacity to pay attention to what is before him. To return to Vico's dictum, man is indeed the great world-builder—but all the worlds of man are located in an encompassing world, that of nature, which is *not* of man's making. Man's own organism and the constitution it imposes on all his activity is one part of this encompassing world; the other, of course, is the natural environment and its structures. Thus, for instance, man's world-building takes place in an environment that has a specific climate and in which specific resources are scarce.

Whatever one might in the end be able to say about constants generated by a "human nature," there certainly are constants that spring from the "human condition." In the language of phenomenology, man is always *in a situation;* his history and his social activity, including identity productions, are likewise *situated*. The aforementioned constants derive from this fundamental fact. They have sociological implications. To wit, these constants eventuate in some general mandates for the social process in which identities can be produced and maintained (the term "mandate" is, of course, not meant to suggest some sort of hidden teleology in society, but the sense of general conditions or, if you will, recipes for social activity). Before turning to the topic of modern identity, it makes sense to look briefly at three of these mandates—not necessarily the only ones, but very relevant for this topic. The mandates are those of order, of continuity, and of triviality.

One of the fundamental mandates, probably *the* fundamental mandate, for any human society is order. American sociology, having been strongly influenced by the democratic ideology of the "American Creed," tended to lose sight of this in a theoretically misleading dichotomization of "consensus" and "control." Incursions of New Left ideology into sociology, with a tendency to equate all order with "repression," haven't helped to clarify matters either. It was, above all, the great achievement of Emile Durkheim and the French school of sociology to have understood that the underlying unity of society is provided by an order of consciousness, an order that Durkheim called "collective consciousness" and (unfortunately, I think) "religion." I would prefer the term "symbolic universe" (as Thomas Luckmann and I have elaborated it in our book *The Social Construction of Reality*), but this terminological change doesn't affect the basic Durkheimian insight. Orderly social processes are only possible through collective participation in symbols. (George Herbert Mead, of course, understood this in a very profound way.) The final order of a society is provided by a coherent, over-arching organization of symbols, providing a meaningful world for individual biographies to unroll in. The term "symbolic universe" refers to this fundamental fact about society—that every society must provide a world for its members to live in. The individual can orient himself in this world, understand his fellow men, undertake projects —in sum, can make sense of his life.

It is important to understand, too, that identity is always located, situated, in a socially constructed world or "symbolic universe." Society assigns identities within the coordinates of its specific world. Thus I am assigned, say, the tag "male" in a world that ascribes very specific meanings to maleness; the tag "teenager" in a world that has inserted the biographical stage of "youth" between childhood and maturity; the tag "with it" in a world (or, in this case, subworld) that bifurcates in terms of "swingers" and "straights" —and, of course, each identity assignment places me in a particular sector of this world's inhabitants, since I'm not the only male, teenage swinger. Conversely, the individual appropriates identities within the coordinates of the same socially constructed world, even when he rebels against the identities he has been assigned. Thus I can decide to reject the behavioral, emotional, and ideological elements of my assigned maleness, and I can try to redefine myself sexually in deviant terms, but even in this process of rejection and redefinition I must reckon with the encompassing world of my society, which remains my situational starting-point, my powerful enemy, and, in all likelihood, my persistent frame of reference. What is more, if I am to have any hope of successfully carrying off my deviant identity definition, I must find at least a few other individuals to give me social support; that is, I must go about the job of rounding up fellow-inhabitants of what will then become a subworld, a counter-society, a social shield against all those others who fail to see the justice of my claims.

The mandate of continuity is grounded in the fact that every society has a history, *must* have a history, since it is impossible for one generation to construct *in toto* the world in which it is to live. We come into the world as children, presented with an endless number of accomplished facts bequeathed to us by our parents, and most of us face the problem of having to hand on some sort of viable world for our children to inhabit. To be sure, we can engage in some dismantling of the parental world and we can try to inflict on our children some favorite improvisations of our own making, but there are, I am afraid, rather severe limits to what is feasible at either end. A moment's reflection about language, which is the most basic ingredient of any social world, will make this clear. We are born into one, continuing stream of language, which contains the sedimented constructions of the past and very largely determines what we or our successors can reconstruct in the future. Maurice Halbwachs, probably Durkheim's greatest disciple, summed this up succinctly, if a little offensively, when he proposed (in his *Les cadres sociaux de la mémoire*) that "society *is* a memory." Mead (without knowing Halbwachs, I should think) links this insight with identity when he suggests that the "I" (the spontaneous, appropriating aspect of the self) is a "figure of memory." In other words, identity (both as objectively assigned and as subjectively appropriated) not only locates the individual in society but also in history. It links him to others so identified in the past and, usually, in the future as well.

The mandate of triviality is rooted in man's deplorable inability to pay attention to too many things at once—which, in turn, must be seen in the light of his biological constitution as an "instinctually underprivileged being" (to use a phrase of Arnold Gehlen, who systematized this insight in his theory of institutions, especially in his book *Urmensch und Spaetkultur*). Because man's instinctual equipment is too impoverished to provide a stable context for his behavior, institutions have to do this for him; that is, they have to provide programs of action that the individual can follow without expending much energy in reflection. This can only be done by trivializing the meanings embodied in social activity, making these meanings routine, taken-for-granted, undeserving of profound attention. Only when this stable "background" is established for his social activity can the individual gain the necessary leeway, from time to time, to have extraordinary, astonishing, attention-provoking experiences. This is true of identity as much as of any other socially constructed symbol or meaning. In order for social life to proceed with any degree of efficiency, identities must be trivialized: "there goes another male—so what else is new?"—"oh yes, he's also a sociology professor." No raised eyebrows provoked by either announcement. The same trivialization is reiterated subjectively; that is, most of the time the individual carries his identity easily, without strain, and definitely without reflection. This trivialization of identity is operative (in Alfred Schutz's phrase) "until further notice"; until, for some reason, its taken-for-granted character is put in question and it becomes "a problem," a surprise, an occasion for reflection or even redefinition. Then, possibly, some admiring female might decide that I am the most routine-shattering male in sight—or, in a more unhappy moment, I might question whether my being a sociology professor makes any sense at all.

I have spent some time on these general theoretical considerations, because they provide a frame of reference within which the particular problems of modern identity can be more successfully negotiated. We can now turn to the question of the peculiarities of this modern identity, as these have emerged in the perspective of sociology. I shall limit myself here to four features that I consider to be important, without pretending that these exhaustively cover the phenomenon:

Modern identity is peculiarly open. This is the quality that David Riesman has pointed to with his category of "other-direction," which Daniel Lerner (in *The Passing of Traditional Society*) has very ingeniously applied to the personality changes that accompany modernization. The openness of modern identity can best be seen in comparison with premodern antecedents. Take the type of peasant society that Lerner was studying in the Middle East. Here, the subjective world of the individual is linked to his immediate social situation in a very narrow and constraining way. This, of course, is true of his range of experience and knowledge. But, more importantly, it is true of his subjective identity. The individual is, as it were, fully encased in his identity. Put more theoretically, there is very high symmetry between objective and subjective identity. As a result, as Lerner found, the individual finds it very difficult to put himself in another's position: he is *what* he is, *where* he is, and even his imagination is strongly immobilized by this location. By contrast, modernization entails a "mobilization" of identity. The individual becomes much more open to new experiences, including new experiences of himself. His horizons expand and he becomes able, first in the imagination, to place himself in quite novel situations. Looked at from the viewpoint of premodern values, this means a weakening of character, a disintegration of the stable structures of the self—and, valuations apart, that is an empirically viable description. As a result, the individual can no longer rely on his original identity, on what he used to consider as his "true self," and must instead rely increasingly on the shifting identity assignments of others in his immediate (and, to boot, continually changing) social milieu. He must shift, that is, from "inner-direction" to "other-direction." Viewed positively, what we see here is an open, adventurous, "creative" personality, which, viewed with a more jaundiced eye, could also be described as shiftless, opportunistic, and lacking in integrity.

But (perhaps fortunately) we are not dealing in ethics here. Rather, we are concerned with empirical causes and consequences. The former we would seek, in the main, in the macrosocial processes of mobility and pluralism, both endemic to modern social structures. Modern societies introduce "movement" of all sorts. People begin to move in space, quite a lot of them physically, most of the others in the imagination. People also begin to move in society, "up" and "down" in terms of status, but also horizontally through increasingly variegated societal spheres. With all this mobility, it becomes increasingly difficult to find closed, self-assured, undisturbed social milieus. Society becomes increasingly pluralistic, in the sense of becoming hospitable to competing and highly discrepant definitions of reality. If we now recall what was said earlier about the conditions for stable identity, it will be clear that what are produced here are precisely the conditions for *unstable* identity. The macrosocial processes of mobility and pluralism are experienced by the individual in his own biography as constant change, uncertainty, and conflict in the definitions of world and self presented to him. More and more, this is true of even very early socialization—just think of the variety of stimuli and expectations that even very young children are exposed to today. This means that the question "who am I?" can increasingly be answered only *hic et nunc:* "I am such-and-such—right now, in this social situation, in this phase of my career"—and, by implication, I may be something vastly different tomorrow or as soon as I switch to another circle of Meadian "significant others."

This, as we may call it, "convertible" quality of modern identity has been aptly caught in Robert Merton's phrase "anticipatory socialization." The modern individual, Merton suggests, is not only socialized into one particular group but, in anticipation, into other groups that he is later expected to move into. But Merton's phrase is too attached to the special case of social mobility in an American-type class system; it doesn't do full justice to the scope of the phenomenon. Modern identity is not only open, it is peculiarly *open-ended*. Modern man, it seems, has an intrinsically unstable identity, never knows for sure just who he is, and thus is ever in search of himself. The reasons for this are not at all mysterious, but are fully available to sociological analysis: they are to be sought, *not* in some strange fall from grace of modern consciousness, but in modern society's social structures to which this consciousness refers. Modern identity is open-ended, "convertible," inherently unstable. It is guaranteed to be in ongoing crisis. This brings us to a key proposition: *the contemporary identity crisis is not an accidental or transitory phenomenon. Rather, it is to be seen as an intrinsic feature of modern social life.* Some further considerations of modern identity may help to flesh out this proposition.

Modern identity is peculiarly differentiated. To some extent, this is already implied in the foregoing feature: the pluralization of social structure will inevitably be reflected by a pluralization of identity. But there is more to it than that. A great expansion and increase in importance of the entire realm of subjectivity are taking place. Arnold Gehlen has called this process "subjectivization" and has traced it, for example, in the rise of the novel as a peculiarly modern literary form. Modern society brings about a far-reaching reality-loss on the part of the institutional order and, conversely, a reality-gain for subjectivity. Modern institutions, for the reasons just mentioned, lack stability. Consequently, they fail to present themselves to the individual as firm, reliable contexts for his activity. Modern institutions are experienced as ever-changing, opaque, unsafe—in the final case, as devoid of reality. In terms of our previous theoretical considerations, indeed, we may say that the underlying reason for this is a consistent violation in modern society of the three mandates of order, continuity, and triviality. The violation of order and continuity should be clear by now; the violation of triviality is just as important. We have suggested that the trivialization of social experience is necessary so as to protect the individual from being ongoingly surprised and thus paralyzed from acting. It is precisely this process of trivialization that is impeded by the instability of modern institutions. They fail to protect the individual from all those surprises. They constantly put him in situations in which he faces astonishing, and at least potentially paralyzing, innovations in the fabric of social experience. They flood him with stimuli (as Gehlen puts it), and with contradictory expectations. One result of this is that the individual becomes extremely nervous; psychiatrically, it can be said that modern institutions are highly pathogenic. But another result is that, inevitably, the individual is thrown back upon his own subjective resources for the stability he needs to live, as these resources are less and less available to him in the objective institutional order. In the extreme case, nothing is reliable or real except the self.

The reality-loss of the institutional order, and the concomitant reality-gain of subjectivity, have brought about a curious reversal in the traditional relationship between subjective identity and its institutionally assigned roles. Traditionally, these roles expressed most fully what the individual conceived to be his "true self." Institutions, institutional roles, and the identity that objectively belonged with them—these were the *realissima* of subjective consciousness as well. Thus, a primitive putting on the mask of his assigned part in the societal drama was *not* hiding himself; on the contrary, he was putting on his real self. Modern man, by contrast, views his institutional roles, and thus his objectively assigned identity, precisely as a mask that *hides* what he "really" is—as, if you will, "alienation," "false consciousness," or "bad faith." The *realissimum* now is his own subjectivity and whatever identity he can make plausible to himself, using the resources of that subjectivity. It may be added that these resources are typically meager, so that the consciousness in question is typically not a very happy one. Needless to say, we have only touched here on an exceedingly complex phenomenon, which requires much greater elaboration. Hopefully, however, enough has been said to indicate a shattering transformation in the construction of identity.

Modern identity is, therefore, peculiarly differentiated, stratified, "interesting." Because the real self is no longer readily at hand in the institutional order of society, the individual is forever inclined to stare with fascination into his own subjectivity and into that of others. Subjectivity is credited with hitherto unsuspected depths and complications. We can do no more than mention here the consistently subjectivity-enhancing trend of modern philosophy and literature, eventuating in the flowering of an intellectual discipline that aptly enough called itself "depth psychology" ("*what* depths?" a premodern observer would probably exclaim). Similarly, we can only mention the implications of all this for the phenomenology of modern emotional life—as in the development of the idea of romantic love, in the new ethos of education, and in the proliferation of "self-discovery" cults and ideologies. It is very important to stress the parallelism between this new interest in the alleged depths of the self and the "flattening out" of institutions in consciousness. The self becomes "interesting" as the institutional order fades in plausibility. Put differently, "alienation" is not only the price but the necessary condition for "self-discovery."

Modern identity is peculiarly reflective. Helmut Schelsky, who was strongly influenced by Gehlen, called this feature "permanent reflectiveness" (*Dauerreflektion*). It means a pervasive propensity to reflect about everything one is doing and, finally, about what one is. Put differently, modern subjectivity is peculiarly reflective and self-reflective. Everything, including identity, is ongoingly scrutinized, explained, brought into full awareness. The phenomenon that Max Weber called "rationalization" is, of course, related to this, though not quite coextensive with it. The macrosocial roots of this are generally sought in the necessary rationality of modern science and technology on the one hand, and of modern bureaucracy on the other. I am quite sure that this linkage makes sense. However, I think that pluralism is, once again, an important additional factor—pluralism in the sense used above, of the pluralization of social worlds in which the modern individual lives. Why? For the simple reason that the individual is *compelled* to become reflective when he is confronted with discrepant definitions of reality. As long as it is possible to live in closed, highly coherent worlds, it is only on rare occasions and in the case of rather few individuals that the official definitions of reality (including the officially assigned identities) become problematic. When, however, conflicting and competing worlds coexist in the experience of the individual, and force him to make certain choices between them, he must willy-nilly start giving some attention to the definitions of reality at issue. In terms of identity, as long as every significant fellow man I encounter agrees that I am *A*, I can afford to take my *A*-ness happily for granted—unless I am an intellectual or otherwise maladjusted. But when some significant others treat me as *A*, some as *B*, and perhaps even some as *C*, it will be hard for me to refrain from devoting some thought to the topic of my identity, however uncongenial such reflections may be to my (let us assume) robust temperament.

This self-critical propensity of modern man has led to a luxuriant growth of therapeutic agencies, whose mission it is to assist the individual in his efforts at self-examination. These are not our concern here. But the same feature adds to the intrinsic instability we have just discussed, and must therefore be seen as an additional cause of the permanent identity crisis of modern man. Although this insight is deeply offensive to all intellectuals, it is, I believe, a simple truth that happiness is commonly associated with unreflecting certitude. A man who, say, is calmly certain that he loves his wife feels little inclination to reflect about this. It is when, for whatever reason, this happy tranquillity is disturbed that he begins to raise questions: "Why do I love her?", "Do I love her now as I once did?", "Do I *really* love her?", and so forth. Conversely, raising such questions is very likely to disturb whatever tranquillity there was before. The same, of course, applies to thinking about oneself: "Am I really *A*?", "Have I perhaps been kidding myself, hiding my *B*-ness from myself?", "Do I *like* being *A*?", and so on. Reflection is dangerous to either individual or collective tranquillity (another way of reiterating the old adage that thinking hurts). Further, reflection is likely to paralyze spontaneous action—as Hamlet knew, and as illustrated in the classical joke about the man who suffered from insomnia from the day on which he was asked whether he slept with his beard above or below the blanket. Thus, the contemporary identity crisis is kept going by a large, probably growing number of people who are chronically addicted to examining themselves.

Modern identity is peculiarly individuated. There can be little doubt that, whatever else it has done, modern society has produced a pervasive ethos of individuality and individual rights. Thomas Luckmann *(The Invisible Religion)* has plausibly suggested that the notion of individual autonomy occupies a preeminent place in the value system (or, as he would say, the "religion") of modernity. It should be emphasized, however, that this individualism has a strongly humanitarian and ethical tinge to it; it is not, for instance, the individualism of the social Darwinist or the Nietzschean loner; rather, it is an individualism in which the emphasis is on respect for the individual and his imputed rights. Further, while all ethical notions have but a tentative relation to social reality, modern individualism is not a *Weltanschauung* floating around in some Platonic vacuum; rather, it refers to an empirical social-psychological fact, that of an identity that is highly individuated in comparison with most previous historical societies. This individuation, of course, can be positively viewed in terms of emancipation, progressive liberation, achievement of "authenticity" (that, once again, would be the properly modern way of looking at it, *à la* Hegel). It can also be viewed as progressive isolation and despair, *à la* Kierkegaard, or, in sociological terms, as a progressive entrapment in what Durkheim called *anomie*—rootlessness, normlessness, separation from meaningful ties with others. We are not called upon here to choose between these two perspectives; *descriptively,* both are correct to a great degree.

The macrosocial sources of this individuation are, in all likelihood, the ones commonly cited—urbanization, capitalism, democracy; negatively, the breakdown of all the old solidarities given in feudalism and still largely retained in the *ancien régime* that followed it. The aforementioned forces of mobility and pluralism, as well as the deeply running current of secularization tearing man out of the security of a divinely ordered universe, must also be considered in this connection. However, the dimension of what one might call (not at all pejoratively) the "softness" of this new individualism, namely its humanitarian-ethical character, requires the introduction of another causal factor. This, in my opinion, is the revolution in the structure of childhood.

To my knowledge, the best scholarly work on this subject is Philippe Ariès's *Centuries of Childhood*. Ariès has masterfully drawn for us the stages in this process of transformation, at least for French society. Beginning with the bourgeoisie, a startlingly new ethos of childhood came to be diffused throughout western societies. Its macrosocial roots are the same as those of the bourgeoisie that was its original "carrier"; specifically, the separation of the family from the processes of economic production and its institutionalization as a protected enclave of "private" life. More recently, though, the revolution of childhood has been powerfully accelerated by the additional factor of modern medicine, which has brought about a historically unprecedented decline in child mortality and morbidity. The end result of all this can, I think, be described quite simply: modern childhood is happier than childhood has ever been before. We, along with only very few generations before us (how many depends on which country one is talking about), are the first human beings since the beginning of history who, when we become parents, can have a reasonable expectation that our children will grow up to adulthood. I think that this is a fact of staggering importance, which, strangely enough, has barely been commented on in social-scientific literature. Its major psychological effect has been that modern parents are emotionally free to invest love in their children from the moment of birth, without having to expect realistically that their imminent grief is going to be all the more bitter in consequence. No wonder, then, that we live in a "child-centered" age! Some pedagogues and psychologists have deplored this fact, and I am not interested in making a defense here of the whole childhood ideology as described, for instance, in John Seeley's studies of suburbia. At the same time, it is important to become acquainted with the physical and social brutality of childhood in earlier periods of western history or, for that matter, in many non-Western societies today, if one is to appreciate the revolution that has taken place. And I will permit myself the observation that one will then find it very difficult to deplore this revolution too seriously—especially if one is a parent!

The individuation of modern identity accentuates the identity crisis in obvious ways, insofar as it makes much more difficult the smooth integration of the individual in the institutional programs to which he is assigned. The "softness" of this same identity, its high expectations of love and of humanitarian concern, accentuates the crisis, especially in the biographical stage of youth. I think that this, again, is an exceedingly important fact, and I regret that I can only refer to it here with utmost brevity. Suffice it to say the following: youth, as we know it today, is a matter of social definition rather than biological fact. Modern society "invented" youth, as an interstitial phase between childhood and maturity; it did so for reasons that we cannot go into here, which are not at all mysterious, but which are fully susceptible to sociological comprehension. Youth has become the locus of intense identity crisis because it is the biographical phase of transition between the "softness" of modern childhood and the (inevitable) "hardness" of the major institutions of adult life. Specifically, the crisis habitually explodes at the point where the young individual first confronts this or that institutional manifestation of bureaucracy (usually, of course, within the institutions of the educational establishment). We raise children in an atmosphere of intense affection and of respect for their individual rights (emotional rights, if one can speak of such, being emphatically included). We then hand over the same children to large, bureaucratically administered, educational institutions, in which, no matter how benevolent these institutions may be, they are "treated as numbers." The predictable result is an eruption of rage. It is very important to understand that the very facts of bureaucratic rationality, anonymity, and utilitarian orientation will provoke this rage, even if these aspects are unaccompanied by this or that offense against humanitarian morality. The quasi-Freudian interpretations of the contemporary youth rebellion (I might cite Lewis Feuer as an example) are thus, in my opinion, far off the mark. Contemporary youth is precisely *not* rebelling against parental figures; on the contrary, it is rebelling against the absence of parental solicitude in the institutional order of society. There are, then, specific features of modern identity that not only ensure a continuing crisis, but make it very probable that this crisis will above all manifest itself among the young.

What prognoses are possible as a result of this analysis? If our analysis is correct in tracing the contemporary identity crisis to deep-seated structures of modern society, we will be hesitant to predict any sharp reversal. More likely, we will be faced for a long time to come with the essential features delineated above. On the other hand, especially among the youth, there have been significant signs in recent years of what can only be described as a massive uprising against modernity. To the extent that many of the structures of modern society may show themselves to be progressively incapable of survival (a far from impossible scenario), this quest for alternatives to modernity may have certain chances of empirical realization; that is, chances of passing from utopianism to social reality. Ultimately, our prognoses will depend on how we answer the philosophical-anthropological question with which we began: Was Thucydides right or wrong in his idea that there is such a thing as a "human nature" and that all of history moves within its unavoidable parameters? If Thucydides was wrong, then we may expect (be it in hope or in terror) that the "liberation" of modern man from the order of institutions will proceed in a straight line. That is, our grandchildren will be more, not less, free of institutionally assigned identities—more open-ended, more "anomic," more reflective, and so on. If, however, Thucydides was right, then there are definite limits to this rupture in the symmetry between self and society, between subjective and objective identity. In that case, sooner or later, there will be a return to the haven of institutions, a return to the notion (and, empirically, the reality) of a society that will be not an "alienation" but a home of the self. In that case, the overriding question will be what sort of institutions these will be—an ethical and political question as well as a scientific one.

Everyday Life and Social Identity
Kenneth B. Clark
College of the City of New York

Educator and psychologist. Founder of the Northside Center for Child Development and Harlem Youth Opportunities Unlimited, and a motivating force in the struggle for black equality. Professor of psychology, College of the City of New York, since 1960. Extensive government service, especially on commissions dealing with civil rights. Author of Dark Ghetto *(1965),* A Relevant War Against Poverty *(1968),* The Pathos of Power *(1974).*

"Everyday Life and Social Identity" is a title which suggests so all-encompassing a theme that one might properly ask by what right does a social psychologist dare to address himself to such a topic?

Essentially, psychology is that science and perspective which is basic in systematic attempts to understand the nature of man. Psychology is the arrogance inherent in the human intellect whereby man validates his claim to uniqueness by his ability to ask questions and his insistence upon demanding or creating answers. And through psychology man may become aware—or block his awareness—of the anxieties and fears and frustrations inherent in the possibility that his questions and his answers may not always reflect the compelling realities of his existence. Psychology is the convergence of the totality of ideas through which man seeks to deify himself, justify his uniqueness, his sacredness—and which at the same time mocks and taunts him and drives him to seek a variety of ways of resolving his conflicts and affirming his being. It is the core of the social sciences—history, economics, political science, sociology. So, too, the significance of both biological and physical sciences must be understood in terms of the manifestations of the awesome complexity and pathos of the human brain and its insatiable quest for understanding of the perceivable and conceivable environment and for the use of these understandings as instruments for control—and as attempts to reduce threats to human life and to enhance the comfort, the creativity, and the positive substance of human existence.

Psychology also can be viewed as the unifying, the cohesive factor which gives meaning to the humanities. Psychology is at the core of religion in the struggle for moral understanding; it is central to the struggle of philosophy for rational understanding and control. It is inextricably a part of the struggle of literature to probe and grasp and communicate the many dimensions of human motivation, both in their frailties and their grandeur. In poetry and the arts psychology is implicit in the experiments in communicating those aspects of human feelings and experience that are difficult to translate into mere words. To the extent that the humanities have their origin in man's inner being, in his attempts to objectify the initially vague yearnings, fears, doubts, hopes, exultations, to that extent they share an inescapable involvement with psychology. In art, architecture, and engineering, in the planning and building of villages and cities and nations—and in the dreams of new towns and utopias—one finds the objectification of human ideas and aspirations and hence another manifestation of the ubiquitousness of psychology.

These are the credentials through which the social psychologist dares the arrogance of considering the problems of everyday life and social identity. Everyday life is the culmination of the totality of human concern and creativity; social identity is the internalization and inner continuity of the impact of the externals of everyday life. The essence of the human predicament is the fact that the human ego—personal identity—has no indigenous substantive ingredients. The self is determined by what is external—what is experienced—and how the environment is experienced and can be manipulated by the individual in meeting his needs.

If one accepts this perspective of psychology as the dynamic source of all human experience, the basis of human curiosity and creativity and hence of intelligence, and of the insatiable cycle of self-perpetuation as these are given reality by continuous introspection and action, one may then most economically define the field of psychology itself as that science which seeks a systematic understanding of man in dynamic interaction with the forces of his environment. Within this simple definition is the justification—indeed, the obligation—for the assertion that psychologists must concern themselves with the totality of man. They must deal with the biology of man—man the animal; the society of man—man in the many complexities of his relationship with his fellow human beings; the technology of man—man the victim of forces beyond his control, and the architect and engineer seeking to master those forces which he believes he does or can control; the religion of man—man the god, and man and his gods; the dreams of man—man the lonely being, man the gregarious being, man the creative being, man in desperate quest of inner and outer substance to give meaning and substance to the illusive chimera that is the human ego.

It is this fundamental theme of the illusiveness of the human ego—its gossamerlike quality, the absence of tangible substance—which I will use in trying to give some coherence to this discussion of everyday life and social identity. Indeed, "social and personal identity" is another name for "ego." Both terms are manifestations of the intangible stuff of consciousness. Consciousness is an artifact—the most important artifact of the universe—a consequence of the unique evolutionary complexity of the structure and function and interaction of the billions of cortical cells which comprise the human brain. But the fact that human identity and consciousness—that is the human ego—are mere manifestations of biochemical intra- and interactions of cortical cells, however complex, is threatening and intolerable and therefore must be denied. It is ironic that the cortical cells themselves demand this denial. Man's brain demands that man sustain a compensatory sense of his own arrogance and grandeur and solidity by asserting his God-like, God-creating, God-obeying, God-defying characteristics. Man's cortical cells insist that he build social, philosophical and political systems that proclaim his superiority and primacy over all other systems of matter and energy—living or inanimate—in the universe. Indeed, they lead him to build elaborate systems of rationalizations to "prove" his superiority over all other men. And they demand that he insist upon his immortality in spite of overwhelming empirical evidence to the contrary.

Man's concrete and ideational creations are themselves the most persuasive support for his insistence that his ego and identity are precious and of substance, and are worthy of enduring respect. Man transforms his ideas and hopes into things. He controls and manipulates his environment to meet his biological and ideational needs and to control his fears and anxieties. He fashions tools and weapons, builds shelters and temples and neighborhoods and cities. He organizes armies and hierarchies to preserve what he has built and plans to build. He has created a civilization of ideas and things to give substance and immortality to his fragile and mortal ego. This is the grandeur—and this is the pathos—of man.

The Irony of Success

The culmination of man's quest for substance and meaning is what he calls success, and the continuity, communication, and expansion of these successes are what he calls civilization. He has achieved scientific and technological triumphs. Today, his material civilization has resulted in a standard of living for more people in European countries and in America that is higher than that enjoyed by aristocrats a century ago. He has built great cities. Man has landed on the moon. He has unlocked the energy of the atom's nucleus and harnessed the sources of power of the earth's resources. He has conquered most of the determinants of disease. He has created instruments for instant communication and rapid travel across the earth and the distances of space.

If such success and this type of technological civilization could give substance to the human ego, the problem of man —his anxieties and gnawing doubts and tensions and conflicts and cruelty and violence—would now have been resolved. Man would have achieved calm and inner security and a stable sense of identity. He would have built a society of peace and justice and beauty. His everyday life would be a reflection of the poetry of contentment—hopefully, a creative and dynamic contentment not a semistagnation.

There is justification in the description of contemporary man as being afflicted by the frustration of Midas—the intolerable despair of a cloying seeming affluence. The Midas syndrome of contemporary civilization has unmasked the practical joker who designed the human predicament to delay the mocking revelations: that only the successful man, who has mastered the more concrete demands and challenges of his environment, can really know the depths of human despair, meaninglessness, and frustration; only the powerful nations, whose power is based upon material and military resources, can know the stark meaning of the loneliness of pomp, the emptiness of bombast, and the impotence of power. Only the affluent can understand the deprivation —the cruel and persistent hunger of affluence.

The promises of material and ideological, democratic and socialist utopias have almost been fulfilled. It is because of this that such terms as "identity crises," "alienation," "existential ennui" have become fashionable trite phrases in the lexicon of sophisticated contemporary discourse. "Urban unrest," "riots," "rebellions," "anticolonial movements," "black liberation," "women's liberation," "student liberation" are all movements which add to the cacophony of anguished outcries of the deeply troubled masses of human beings throughout the world. Their cries reflect their anger that the promises of personal and social identity have not been kept by one or another form of utopianism. Ideological contracts have been broken in the process of negotiations. The promises engendered by the end of colonialism, the expanded education for the masses, the production of more telephones, more washing machines, more airplanes, better wages and more leisure for workers, bigger and better libraries, more books and magazines, more and better equipped museums have not been fulfilled.

In achieving the *things* which were supposed to resolve the identity crisis, that crisis has emerged more starkly and more directly threatening. Is it possible that in achieving *things,* we removed a basis for identity which was inherent in the quest? Success in attaining the means, only to find that the ends are, or appear to be, even more remote, is a mocking psychological joke that men who were prepared to believe in ideologies and utopias, and who were educated to hope and to work to make dreams real, were not prepared to endure.

The restlessness and rebellions of nonwhites, youth, women, and the oppressed throughout the world cannot now be understood in terms of the facts of the persistence of human cruelty, injustice, and the varieties of subtle and flagrant oppressions. These are real and are intolerable to sensitive, moral, and concerned men, and will continue to motivate them to seek positive solutions. But cruelty, injustice, and oppression have long been a part of human societies. In fact, one might make a persuasive case that they are less flagrant and barbaric now than they were even a century ago. Then, they were more commonplace—or God-given; now they are more disturbing because we have been taught to believe that they are remediable—that man can control his environment and make it more conducive to human aspirations. This tantalizing belief has been reinforced by mounting, staggering evidences of success in man's ability to master his material and biological environment. But the joke becomes all the more tragic because the more grand the successes, the more clear it becomes that they exist in the context of many dimensions of failure.

—Man's landing on the moon must be seen as part of the continued dehumanizing filth of the slums.

—Instruments of instantaneous communication coexist with the inability of man to communicate across international, racial, religious, sex, and age gaps.

—Monumental cities of glass and concrete defy elementary taste in esthetics and respect for the important intangibles of humanity.

Kenneth B. Clark

—Scientific and technological breakthroughs which are the products of mass education make more unacceptable the fact of illiteracy anywhere in the world—and make reprehensible the fact that the public schools in the largest cities of the most affluent nation of the world spawn hundreds of thousands of functional illiterates yearly.

—The increased efficiency in the production and distribution of food makes the fact of hunger and poverty anywhere more intolerable and guilt-producing for those who are too well fed.

The phenomenon of contrast between the successes of our civilization and the persistent failures of slums, ghettoes, racism, ignorance, poverty, and hunger is made all the more intolerable to the sensitive human being because these discrepancies not only exist, but by virtue of science and technology can be instantly and seeringly communicated. We knew that Harlem and Biafra and Vietnam and the other human cesspools of Africa and Asia and South America were an abomination because our television and radios and newspapers and magazines told us so instantly. We were forced to be accessories to the horror of one human being coldly murdering another human being because our TV brought this into the privacy of our homes.

This is a major ironic ingredient of civilized everyday life today: the instruments of civilization reveal the lack of civilization; the repeated promises of justice and morality betray the facts of injustice and immorality. Furthermore, these contradictions are brought to our attention immediately, intensely, and inescapably. Modern man—and certainly the youth of these times—no longer are permitted the luxurious and protective escape from and repression of disturbing injustices. Contemporary everyday life burdens us with the fact of having to know that civilized man is capable of being efficiently barbaric, cruel, and inhuman. This is the anguished, identity-destroying burden which the youth throughout the world are seeking to avoid through their rebellions; their "new cultures" and "now" cultures; their defiance and rejection of the values and standards of the past; their ridicule of our ideologies and their mocking of our hypocrisies; and their flirtations with escape through drugs, passivity, and their own forms of inner emptiness. Probably the most frightening consequence of the revolt of the youth—understandable, although not acceptable, in light of the fact that this is the first generation of human beings required to live their entire intellectual lives under the shadow of imminent extinction of mankind—is their rejection of the laws of rationality and the value of values even as they cry out against the violations of reason and justice on the part of their elders. Many of the more sensitive youth at this stage of human history seem to be left without the compass and rudders of a functional sense of ethical values. In fighting against the misuse of these instruments by their elders they seem to be destroying the possibility of using them more effectively and humanely by themselves.

Barriers to a Positive Sense of Identity

In addition to the problems and stresses inherent in the coexistence of affluence and the social and psychological residues of deprivation and their related injustices and inhumanities—and the fact that these are instantly and dramatically communicated—there are other characteristics of contemporary everyday life, probably inevitably associated with an industrial, technological, scientific civilization, which make a positive and morally effective self-identity difficult, if not impossible, to achieve and sustain. Among these are the requirements for reacting to, or avoiding, the persistent forms of injustices; the need to satisfy the demands of efficiency and mass production through reducing the individual to a mere interchangeable part of a standardized collective mass, and the standardization and reduction of esthetic, philosophical, artistic, musical, and literary values and tastes to a norm compatible to that which can be mass produced, packaged, and sold condescendingly, if not cynically, to the masses of rejected human beings. These are more subtle, but no less seering and dehumanizing forms of inhumanity.

Another component of the contemporary version of the human joke is the fact that those who are rebelling against injustices and arbitrary violence are themselves often unjust and violent. Those who are crying out against the consequences of the dogmatism of others are frequently themselves blindly dogmatic. Those who are disturbed by the cruelty of others believe that this justifies their own cruelty. Those who are frustrated by the pattern of inhumane irrationality are often mindlessly irrational in seeking remedy. Probably the most tragic of these everyday ironies is to be found in the fact that otherwise sensitive human beings who are concerned about the insensitivity and violence of others, themselves resort to insensitive violence in seeking a more moral system. It has become a fact of contemporary everyday life, that individuals seek identity by accepting simplistic premises; by accepting dogmatism; by accepting uncritically the assumption that individual human beings are expendable in the interest of some higher cause—that there is some higher morality which justifies transitional immorality. This is a modern, pragmatic morality which is consistent with modern worship of efficiency. There remains the shadow of insight that a morality of efficency is immoral, even when it is the modern fashion.

One way of avoiding even this appearance of moral commitment which contemporary life requires to be efficiently immoral is to refuse to be committed; to accept the injustices of contemporary life as given; to seek one's own isolated peace and personal success; to accept personal achievements and acquisitions as the measure of personal identity; to accept a collective identity as the measure of individual worth. This approach is easy and common and requires relatively less expenditure of energy. But it, too, appears to be a form of passive—and sometimes active—immorality.

In both his rebellion and his passivity, contemporary man seems reduced to standardized parts. His individuality is sacrificed in a collective herd. Mass-produced machines serve mass-produced man. The high price of a higher standard of living made possible by mass production is the subordination of creative individuality and its resulting doubts concerning the worth of self. In a society of affluent and deprived human beings, not differentiated by the absence of inner emptiness, there is the common bond of the necessity to substitute a collective identity for the loss of an individual identity.

In such a society of collective identities, ideas as well as buildings, education and politics, beliefs and what passes as art are constrained by conformity—a deadeningly efficient sameness. Such conformity, in turn, destroys esthetic concerns. The mass production of glass and concrete buildings of affluence, mass-produced slums, the mass production of suburbia, mass transit, the mass-produced pathology of drugs, the ultimate in mass pathology in the collective deaths of atomic warfare wherein the individual is not even permitted the right to individuality in death—all of these are ugly. In leaving no room for individuality, they leave no room for creativity, beauty, compassion, kindness, empathy, and love. This type of contemporary worship of efficiency, demanding as it does a collective identity, destroys the possibility of an individual identity. This destruction of the individual's creativity perpetuates and intensifies personal and social destructiveness. It intensifies tribalism, parochialism, nationalism, and racism. It makes difficult an affirmative sense of self as an individual worthy of respect because he is an individual—and makes impossible respect for the humanity of other human beings.

It is a truism, embarrassing to assert, that only human beings who lack respect for self and others could permit slums and ghettoes to exist when they are correctable; could permit human beings to live in houses and neighborhoods of filth and ugliness when this is correctable; could permit generation after generation of human beings to be destroyed by an indifferent and inefficient education system; could permit a soulless higher education and inner city illiteracy; could permit our art, literature, and media to become so bound to commercial determinants of worth as to become instruments of contempt for man and symbols of the hollowness and mockery of the human soul; could permit disease, death of the body and the soul in the midst of the knowledge which could heal; could mouth the importance of environmental conservation while the critical problem of human conservation is ignored in the face of the human beings whose humanity is permitted to be wasted.

These are not paradoxes. These are the realities of everyday life and these are the profound problems of personal and social identity. They are so inextricable that the negatives of the one determine the negatives of the other—and if these negatives are not reversed they will increase geometrically.

Affirmative solutions to these problems may no longer be possible. The interrelated components of the barriers to an affirmative and creative self may not be removable as long as they are an integral part of the complex system of arrangements and benefits which we call modern civilization. And certainly it is not realistic to believe or to hope that the "progress" associated with civilization will be reversed merely to permit contemporary man to try for some other approach to meaning and substance for his life. Somehow, however, the incorrigible propensity toward optimism, wishful utopianism will continue to propel some·men toward seeking positive and humane answers to human problems.

Following in this tradition I suggest the following—not for rational but for affective and hopeful consideration: if there is to be the affirmation essential for a positive sense of personal and social identity, somehow our instruments of technology and communication, and transportation and engineering and education must be mobilized and directed toward persuading the masses of human beings throughout the world that the human brain has made it possible to assure everyone the basic essentials of food, health, shelter, a clean community. Hunger and squalor can, and therefore must, be eliminated. This is a primary step toward regaining respect for human beings. This is an inescapable requirement of human dignity. We must understand that some of the same methods that have led to the partial successes we call civilization can be utilized to achieve a truly civilized society.

Such a transformation of means is essential if the perception of everyday life is to be changed from negation and guilt to affirmation and esteem: again, paradoxically, we must move toward building within human beings a sense of their own finiteness, their own limitation, as this is an integral part of human positives. Ironically, the fulfillment of man's potential depends upon his acceptance of his own finiteness.

We must try to educate ourselves and our fellow human beings to accept the fact that an effective functional identity cannot be guaranteed by the mere manipulation of the outside—the external environment. Acquisition of material things is relatively easy; the internal environment is crucial. After the external requirements of justice and humanity are obtained, or vigorously sought, identity—the creative sense of self—depends upon one's own sense of confidence and reality. A positive identity is dependent upon one's ability to accept the limitations of other human beings as an aspect and reflection of one's own limitations. One, therefore, must be unwilling to settle for the seductive appeals of the dogmas and simplistic explanations and ideologies which make it possible to justify cruelties and death inflicted upon those who do not happen to share our current dogmas.

Somehow, we must try to understand that the ultimate irony may be that there is no "truth" or "good" or "justice" which justifies the denial of the quest for "truth" or "good" —and that there is no "justice" which justifies even temporary injustice—or cruelty.

We must try to understand that while man must compromise in his methods, he cannot stabilize the tenuousness of his own ego—or seek to obscure his guilt and anxieties—by inflicting verbal or military or repressive barbarities upon his fellow man.

Tenuousness of ego, guilt, and anxieties must be affirmed in the very act of accepting the difficulties of affirmation—in accepting the functions and the temporal and intellectual and moral limitations inherent in being human, mobilizing the individual's available resources to engage in the struggle which differentiates dynamic life from inert matter.

In this broadened perspective of the affirmation of the

promises and limitations of the human ego there is and must be an inevitable liberation of the human spirit which might make it possible to accept the realities of life and the reality of death of self—and of others. In accepting these twin realities, and in respecting the identity of others, may be the answer to the taunting question of how to give "substance" to that inherently substanceless but most precious substance, the human ego.

Perhaps the ultimate paradox in seeking an answer to an affirmative sense of self in the contemporary world is the recognition and acceptance of the possibility that social and personal identity can only be affirmed in the context of contemporary life if it is affirmed in spite of the realities of everyday life. Self-affirmation can occur—and probably only occurs—in spite of the many powerful forces which seek to negate the self. The creative identity of the self is possible only when the temptations to seek dependence upon the spurious crutches which promise but mock affirmation are resisted.

Cultural Integrity and Personal Identity: The Communitarian Response

John W. Bennett
Washington University

Anthropologist and educator. Specialist in Japanese culture and society. Professor of anthropology, Washington University, since 1959. Extensive research on societal organization in Japan and related economic and ecological findings. Author (with Iwao Ishino) Paternalism in the Japanese Economy *(1963),* Northern Plainsmen: Adaptive Strategy and Agrarian Life *(1969, Rev. 1971),* The Ecological Transition: Cultural Anthropology's Responsibilities and Knowledge in Human Ecology *(1974).*

After all, the community has never been one great happy family for all men. TOM WOLFE

Social anomie and alienation lead to the disintegration of ties and loyalties, and, eventually, to the loss of personal identity. The large organization of our time breeds alienation and, by thousands, people move into groups, seeking ties in common interests, love, recreation, books, records, drugs, crafts, religion—rupturing the old class lines and social categories as they do so. This return to the group is a central social fact of our time, and however futile or superficial it may appear, there is no doubt that it reflects a sickness in the social system. Vine Deloria, Jr. wants to rebuild modern society on the flight; he advocates legal guarantees of group solidarity.

There is nothing new about this. The search for cultural integrity and personal identity is as old as civilization and, especially, as old as Western civilization, where the problem seems to have been recurrently acute. I focus on one important phase of this search: the communitarian tradition—the oldest "new" and the most traditional "experimental" social movement in the Western world, a persisting template of sharing and interacting that undergoes repeated revival as an alternative to the majority institutions. For twenty centuries, beginning with the Qmran community in the Dead Sea, people in the Western tradition have sought to escape from the tensions of acquisitiveness, amorphous freedom, and social hierarchy, toward the sharing of possessions, decisions, and brotherly love. The movement has repeatedly "failed," as its communities have disintegrated, yet repeatedly "succeeded," as its message has become a prod to the conscience and a hopeful vision of an alternative.

What is this—this communitarian tradition, this communal form of life, this form of community, the commune? It is simply a recurrent pattern of ideals and social arrangements. The ideas are a "cognitive map" of impressive simplicity: (1) share thy possessions; (2) share thy tasks and decisions with others; (3) minimize thy wants; (4) love thy brother, who is all men. From this template, based on early Jewish and Christian ideals, comes, in Georg Simmel's phrase, a "social form," or a set of customs and institutions which vary depending on circumstances and setting, but which generally feature some or all of the following: the collective rearing of children; collective decision-making processes; egalitarian relationships; communal property; sharing work; maximization of interpersonal interaction; many complicated rules of everyday living. As communitarians adapt to the world around them, they also need to remold these institutions and ideals, and this results in what I shall call "paradoxes," or intersections of communitarian principles and practices with the external social reality.

I have defined the mission of the communitarian movement as dual: the search for the integrity of the *gemeinschaft*, on the one hand, and the flowering of the identity of the individual in the group, on the other. But the pathos of the human condition is defined by the paradox hidden in these two goals: the fulfillment of the personality and the consolidation of group culture are not inherently or always compatible, and the achievement of one may negate the other. The communitarian tradition thus is not the solution to the problem of man, though it may help to bear the agony. My major theme is simply that communitarian ideals and their concrete expression—the communal society—are not the heaven for which everyone seeks, but rather a clear demonstration of the need to search for that heaven. The communal group is no paradise; it replicates, in slightly different, often accentuated forms, the basic dilemmas of the individual in society, and even at its most ideal, the commune, or any small group, contains most of the frustrations of human existence. Indeed, the current most enduring forms of the commune—the Hutterian colony and the Israeli kibbutz—have lasted precisely because they have been willing to compromise with the majority institutions and suffer more than a little corruption, or transformation, in the process.

I believe that the most important function of the communitarian ideal and the communal society is to provide a demonstration of the possibility of the capacity of humans, or at least some humans, to live more cooperatively and more humbly. To go all the way in this direction, as I shall suggest in this essay, is difficult and restrictive, and while there seems little chance that human society can be reconstructed along these lines, perhaps it can be influenced in the general direction. Hopefully, also, if we understand the social and emotional costs of communalism, we may be able to construct more realistic social transformations. If man is indeed an evolutionary product of predation, who can live in brotherhood only under unusual circumstances, then this is all the more reason to emphasize brotherhood. The communitarian tradition is a small voice in the wilderness of human barbarism, but things would probably be a lot worse if it had not been there.

There is a matter of terminology to dispose of. I have mentioned the "communitarian tradition" and the "communal society," the first term referring to an ideological template emphasizing sharing and community; the latter referring to an existing social entity exemplifying these ideals in varying ways. This sharing business is indeed the emphasis in this essay. However, there are other terms. One of these is the "millenarian movement." This refers to attempts by people to establish the ideal human society of the distant millennium in the here and now—attempts which have probably featured communitarian ideals more often than not, but which don't have to. A related type is the "revitalization" or "revivalistic" movement. These are generally met with among tribal or former tribal peoples, or any group with ingrown cultures threatened by external power. Much less often than the types classified as "millenarian," do these latter take the form of true communitarian or communal societies. Still another term is "utopian community," referring to any social experiment that seeks to make a social form considered to be outlandish from the point of view of the establishment. Obviously, communitarians are "utopians," if we or they choose to consider their way of life as dreamlike, idealistic, utopian.

Thus, we are concerned with all those millenarian, revitalistic, or utopian movements that have taken specific communal forms. I focus on the communal element because I believe it represents the main goal of the search for alternative social pathways. This throws the analysis into sociological and social-anthropological channels, since communal life requires attention to daily life and social organization. I believe this emphasis to be a good one because nearly all the discussions of the communitarian tradition have been preoccupied with ideals. Since the majority of communal experiments have failed as communities, the focus on ideals results in ambiguous conclusions about the value and effectiveness of these ideals. By shifting the emphasis onto social relations and the day-to-day operations of these communities, it becomes clear that imperfect humans cannot possibly live in complete accordance with the ideals. This directs our interest toward the struggle to be better human beings, and away from the "failure" of idealism.

I.

To bring communitarianism into focus, I want first to review the historical line of development. This begins with the Galilean withdrawal sectarians—the Essenes and others—of the last century or so before Christ, brought back to our consciousness by the Dead Sea Scrolls and the ruins of the Qmran commune, where the dissidents sought to escape the alienation and corruption of Roman Palestine. Jesus' own posture was complex: He certainly agreed with the sectarians' criticisms of the establishment, but He seemingly rejected the withdrawal solution and taught instead a technique of passive-militant coping with the world in an effort to change it. The facts are elusive and no one will ever know for sure, but something in the so-called Judaeo-Christian tradition and its antecedents started this engram of sharing, and Jesus, if He did not actively seek to drop out, did at least preach, in Acts IV, some kind of renunciation of property (the exact meaning remains unclear—the passage can mean either communalism or simply charity and austerity).

The early Christian communities of the first and second centuries A.D. may or may not have been "communes"— here, too, the evidence is hazy, and it is possible that their communalism was more a matter of survival under conditions of privation than of specific ideals—a situation which we will encounter again and again (see, for example, David Plath's translation of Sugihara-san's narrative of the establishment of a Japanese commune, where communalism emerged out of necessity in a group of families who got ostracized from their tight little village; or, for that matter, the history of the kibbutz movement, which arose in part out of trying to farm a desert with few of the tools and none of the capital needed to do the job).

But *something* persisted in this early Christian period, enough to keep the tradition alive down into the twelfth and thirteenth centuries, when the mountain paths and town ghettoes all over Europe were overrun from time to time by wandering fanatics, militants, flower people—most of whom preached brotherhood and poverty and were a thorn in the side of the Church, being treated like heretics and largely buried from sight until Norman Cohn [1] brought them out of the rare book rooms. Some of these groups persisted and "institutionalized" their ways of life, becoming substantial communities like the Waldensians of the Alpine valleys, the Bohemian Brethren, or the Taborites, and influencing the religious movements culminating, in the sixteenth century, in the Lutheran Reformation and the more diffuse Anabaptist movement, or the "radical reformation," as George Williams [2] calls it. The sixteenth century was another period of communitarian flowering, the protests being directed at both the Catholic and Protestant hierarchies and the emerging bourgeois, capitalist order.

Pietism and other forms brought these movements down into the eighteenth century, often decimated by persecution. But something else entered the record in the eighteenth: the rationalism and sociological secularism of the *Philosophes*, and at this point the communitarian tradition divided into the old religious, and the new utopian socialist style of Fourier, Saint-Simon, and others. And then communitarianism moved directly into the nineteenth century, when the wilderness of America offered space and isolation to try it out, and a great many people did: the utopias of the frontier, probably the best documented of all communal experiments.

And so down into the twentieth century. The scene has been dominated by the Hutterites (the only true communal survivors of the sixteenth-century Anabaptist flowering) and the kibbutz, the successful version of the eighteenth-century socialist type. But while these have received all the notices, every decade of the twentieth century in America has known at least a dozen unpublicized communal communities eking out an existence somewhere in the redneck hills of Georgia, the humid woods of Missouri, along the dusty roads of California, or in the green austerity of upper New York State or Pennsylvania. In the twentieth century the movement has spread beyond America and Europe, and into Japan and India, the two Asian countries most deeply influenced by what we are beginning to doubt is Western civilization.

The contemporary North American and European scene is very complicated. First, I acknowledge a powerful generalized movement into small groups of all kinds, the bonds deriving from hobbies, sports (the Pump House Gang), physical handicaps, drug-taking, homosexuality (see Edward Sagarin's account of "societies of deviants"),[3] and, of course, specific communitarian ideas, the whole representing a flight from massness, the status race, and a search for identity in the embrace of some kind of consistent culture. Parts of the picture resemble the Middle Ages—in fact, the hippie-flower children phenomenon is virtually duplicated in the medieval descriptions of itinerant and semi-itinerant communities and

Billy (Dennis Hopper) and Wyatt (Peter Fonda) bid goodbye to Stranger (Luke Askew), center, as they leave the hippie commune in this scene from the motion picture "Easy Rider." Copyright ©1969 by Columbia Pictures.

bands, wandering across the countryside, dropouts from a hierarchical society devoted to war and vainglory and churchly pretense and corruption. But there are also recognizable communities in the contemporary scene, rural and urban; the former approximating the true self-sufficient commune; the urban, usually a more amorphous, experimental type, its members wage-earners, coming and going—more of a residential co-op. Which among these many kinds of groups are we talking about in this paper? All of them, to some extent, but particularly the more sedentary attempts at establishing a settled community. However, it should be clear that the enormous variety of these groups is witness to a malaise in America and in all industrial societies, and I am concerned with this malaise however specific I get with respect to particular responses to it.

Now I wish to discuss some of the main themes of this evolution. First of all, I have implied that the Western world is the homeland. But before this is dealt with in substance, I wish to bring up a difficult issue: monasticism. This is relevant here because both the West and the East have had vigorous and long-lasting monastic institutions, and monasticism is a form of communalism. I believe the issue is clear: most monastic orders in the West originated as attempts to escape the corruption of the established Church, and the communitarian message of poverty and equality was the obvious dialectical countermessage to corruption and hierarchy. The Church, appreciating the dangers, simply co-opted the orders, and they changed from communal experiments into unisexual business establishments with their own wealth, complex hierarchies, and modifications of the doctrine of sharing and equality. In the Orient, similar transitions occurred, particularly in Japan, where the great monastic orders sometimes wore swords and led political riots. In other times and climes, Buddhist monasticism took the form of mendicancy, or of temporary retreats for men in secular life. Certainly many forms of communal social organization can be found in monastic orders, both male and female, and in all honesty I should perhaps include the type. But one has to make choices and I choose to focus on those voluntary communitarian groups that have maintained a clear dissenting, rejecting attitude toward the majority society, and that have preserved the elementary sexual function of breeding and some form of the family.

Something also should be said about the Chinese communes and other agrarian collectives. I intend to exclude all cases of government-planned or compulsory agrarian collectivism: the Chinese communes, the Soviet *kholkhoz*, the *ejidos*, the Taiwan veterans' collectives, and similar institutions. This suggests that spontaneous idealism or at least voluntarism is an important criterion for me—the government collectives are dogmatic experiments, more or less successful, devised to solve problems of food production, land tenure, and restlessness; they are captive communes, like the monasteries. Moreover, most of them are only partly communal, since many pay wages or permit individual shares, provide for private farming, or control the lives of their members with government-enforced sanctions. But, like the monasteries, we have to admit that the agrarian collectives share some of the organizational details and social dynamics of the voluntary communes.

If we identify the true, voluntary, dissenting commune with the West, we have to distinguish between means and ends; specific ideals and how these are to be carried out. At the level of ends or ideals we find close similarities in Oriental and Western dissenting doctrines: the rejection of the pleasures of life, the yearning for austerity, and the need to cast off the distractions of possessions; the emphasis on peace, love, and brotherhood. At this level the similarities between Christianity and Buddhism, Hinduism, Zoroastrianism, and so on are well-known and in the last analysis they are all part of the same milieu: Hither Asia.

But there is a substantial difference in the means advocated to reach these ends: in the East, it was individual withdrawal, the loss of self, recognition of the imperfection of the world, and hence a turning to Heaven, either by a retreat into hermitage or monastery, or by attempting to live it out in daily life. In the West, the search for the perfect world took different forms: toward the discovery of true individual selfhood, rage at the imperfections of the world and an attempt to cure it; withdrawal not into passivity but into little intentional societies that sought to bring the millennium to the here and now; in short, a somewhat more activist set of means, and a dangerous set at that, because there is nothing riskier than actually trying to make dreams of perfection come true.

Thus, while the communitarian ideal protests the majority tendencies in Western life, it is also a rich expression of a very basic feature of Western thought and behavior—the striving for social perfection and mastery over the existential present. This "millennial" element—the idea of actually trying to create the perfect society on earth—is foreign to the traditional Oriental posture, where resignation to the corrupt status quo is the dominant theme, with Nirvana in the afterlife. The gods are largely uninterested in human strivings and hence striving is useless and even objectionable (with the exception of Japan, where the secular, power-centered ethos of the culture from earliest times built an activist, mastery element into both Buddhism and Shintoism). Thus the Grand Illusion of the West—that man can achieve perfection—is at the bottom of both the destructiveness of Western culture and also its most ancient internal critical response: the communitarian tradition.

Within Western civilization it is the dualism of the Judaeo-Christian stream of thought that forms the concrete basis of the communitarian ideal. Conceptions of a perfect heaven and a very imperfect humanity gave rise to the struggle to make the City of Man conform, at least to the best of one's ability, to the City of God. Jesus' own teachings exemplify this tension: on the one hand he held that man was imperfect and corrupt—ideas with an Oriental flavor and similar to notions held by the Essenes and other sectarians, influenced by Persian and Middle-Eastern cults. But on the other hand, Jesus taught that most men had to live in the imperfect world and therefore work to improve it. After Jesus' death, this ideal was reinforced in the myth of the Second Coming, a form of desperate optimism that became the key idea of the millenarial dissent groups of the Middle Ages, most of whom (some of them communal) believed that their mission lay in preparing corrupt humanity for the Coming. Such ideas also blended with classical concepts of the Golden Age; hence, the perfect Christian society should be a replica of the early apostolic Nazarene community of Jerusalem described in the Book of Acts. These ideological odds and ends—the Golden Age, the Kingdom of Heaven, and the Second Coming—combined to form an ideational narcotic, driving group after group to experiment with utopian social and economic formulas.

John W. Bennett

"Social and economic." The point is that any stratified social order, with its existing inequalities and injustice, could be defined as a corruption of Jesus' teachings and hence a "fall" from the true apostolic model. The communitarians did not except the churches: the Catholic, and after the middle of the sixteenth century, the Protestant as well, were considered distortions of the original Nazarene template and hence in need of challenge and reform. These challenges were not simply doctrinal, but involved specific formulas of social and economic polity: in particular, a communalizing or "communizing" of the basic means of production, in which the household or community produced food and goods for the good of all, minimizing their disposal for profit and exploitation. The various historical communal groups differed in the extent of application of these principles and in their attitude toward private property and in techniques of group decision—above all, the assembly of voting equals.

II.

Now, precisely, what are communitarianism and the commune? Communitarianism (the beliefs) and the commune (the living expression of the beliefs) are first and foremost attempts to escape from the majority society. But this escape is a curious and paradoxical thing. For the commune cannot really escape the world; it can only challenge it. To survive, however far one wishes to remain from organized society, requires compromises.

What are these needs for compromise with the world? To farm efficiently, to produce furniture or tools and to use and sell them, to manage estates for the nobility (as did the Hutterites), to sell farm produce on the market, to deal with governments to get land, to deal with tradesmen to get things one cannot make oneself, to deal with banks and moneylenders to get capital—in all of these things and others the communal societies must confront and imitate the very institutions they seek to change. The way of the communitarian community is along a knife-edge, constantly risking a long fall into the worldly abyss, avoiding the worst by strict moral management and endless rationalization, and trying to avoid the slip into authoritarianism and the corruption of the brotherly ideal. It is no wonder the Hutterites say "we do not live this way because we like it"; *i.e.*, Christ's way is the hard way; brotherhood is the hardest task man can set himself to perform.

Such communities are always a minority, never a majority. No large social or political entity has found it possible to keep to the austere and egalitarian path, and in only two or three cases—like the Taborites or the confused Anabaptist faction that seized the city of Münster—did the communitarians fall into control of a "normal" community. The larger society of the West, while grudgingly acknowledging the validity of some of the ideals, has never really come to believe in the system enough to renounce economic growth and its individualizing and disequalitarianizing force. However, the communitarian idea has sometimes been felt rather widely: the Anabaptist impact on the established Baptists and the Fundamental-Pentecostal sects; the utopian socialist communities on the political socialist movement; the broad-ranging inspiration of the Israeli kibbutz (there are now kibbutz-inspired communes in fifteen countries around the globe).

Morally or sociopsychologically, communitarianism espouses a way of life which finds its greatest satisfaction in group participation rather than in the personal fortunes of the individual. In a communal society, people are expected to to do their best for the commune, the group, and not for themselves—although it is also usually assumed that if they do this, they are also reaching their own personal fulfillment. This is a matter of belief, not of psychological necessity, and in recognition of this, communal institutions are usually constructed in order to encourage or require such behavior, and here is the sticky part. How do you "encourage" or "require" in a milieu which may assume that people *know* what is best or else they would not be there? If they do not *know* (believe), then they have to be shown, but if you *show* them you violate, to some degree, the voluntaristic spirit and you risk the emergence of hierarchy or inequality (a major theme in Mary McCarthy's short documentary novel about a New England commune, *The Oasis*). The situation is exactly the same in the economic sphere: obviously not everyone can perform all the tasks, and since some tasks are clearly more important to survival than others, or more pleasant, there is always the risk of jealousy, hierarchy, and emergent "status" (another theme of McCarthy's novel).

And then, of course, there is the paradox of compensation. In the workaday world we say, "a man won't work for nothing," meaning by this that he needs a living wage. The communes have a problem with this since to pay wages is to introduce an individualistic note—a *wage* is a man's own thing, he can do what he wants with it, but since the commune emphasizes share and share alike, it must strive to keep property communally owned, or at least to reduce the individual share to a low level. Hence, you can't really pay *wages*. But it is also true that some form of compensation, human psychology being what it is, is hard to avoid. The socialists formulated the principle as: "From each according to his ability, to each according to his need"—a doctrine that recognizes individual differences and is fundamentally ambiguous as to share-and-share alike.

In fact, communes vary in their procedures for compensation: some, like the Hutterites, pay no wages of any description—the commune furnishes all necessities, plus uniform supplies of personal possessions, plus a few perquisites for the Elders. But even in the most committed of colonies, people accumulate possessions, often secretly. Other communes—especially the near-communes like cooperative farms, the Israeli *moshavim,* and others—may pay equal wages to all. There is no absolute uniformity in custom; and what may be compromised on the compensation front will have to be made up somewhere else, like stricter child socialization. All communes therefore experience a problem over the ideal of shared uniformity and nonacquisitiveness, and the fact that human beings have wants, however thoroughly socialized they may be in the doctrine that wants are a source of evil.

Consider the following list of human activities: manual labor, manufacturing, food preparation, eating, worshipping, raising children, managing possessions, family life, sexual relations, love and courtship, studying. This list is arranged in a rough order of the activities which are most commonly performed communally. Obviously, work of all kinds, property management, food activities, and worshipping are relatively easily done in or by groups, with approximately equal responsibility and sharing, at least for a majority of the participants. On the other hand, family life, sexual experiences and courtship, and studying are not amenable to full or easy communal participation, although communal objectives can influence them, and experiments can be conducted in participative involvement. The raising of children is an example of an activity that stands somewhere in the middle: it can be done by individual nuclear families, by extended kin groups, or by a commune, depending upon belief and specific institutional planning. The child-raising issue is a crucial one since the commune must train its upcoming generations—particularly if it is a new one—in the techniques and ideals of group living.

Clearly, some control over the children needs to be asserted by the commune. But there is a certain area of movement here. The Hutterites have their individual nuclear families live in separate apartments and rear their children up to the age of three, at which time the community school system takes over—although the children continue to sleep at home. On the other hand, some of the early kibbutzniks attempted to eliminate formal marriage and the family completely, rearing the children apart in age groups. But this extreme custom did not endure, and a certain amount of individual family life has reemerged in the kibbutz: children are still educated apart but now come "home" to the famous "family hour," vacations, etc. Despite its compromises, the average kibbutz still goes one step further than the colony in communalizing child rearing, although both systems appear to be equally effective in preparing the children for life in the commune.

John W. Bennett

The point is that communalism is never total—humans are not ants—but must always be adjusted to specific ideologies and to certain regularities in human psychology and social life. Clearly, sexual experience and love have a strong dyadic character—it is impossible to thoroughly communalize them, despite the kibbutz attempt or the myths of "primitive sexual communism." For bioemotional reasons the cathexis is largely limited to two persons at any one time, and while attempts at broadening the participation may succeed for a period, the system will tend to give way to more limited associations, especially when the family issue begins to arise. It is, quite obviously, much easier to *labor* collectively, and so indefinitely, because fewer emotional factors are involved, and because a great many tasks can profitably use more than a dyad.

Moreover, the precise type of communal sharing will vary within the same activity. Consider property: this is another crucial factor because of the acquisitive drive and its individualizing tendencies; hence its management and control always become a key institution in communalism. Here are some of the variants: the property in a commune that is defined as communal or personal can vary in quantity and kind—from minimal-communal to minimal-personal. Property of any kind can be considered to be free for the taking by all members, or it can be considered to be owned by the collectivity and assigned to individuals by vote or by executive decision. Personal property can be permitted to individuals who are allowed to acquire it with their own efforts, or it must be allocated to individuals by the collectivity. Property can be viewed as strictly collective in perpetuity, with individuals or family units having no equity; or, a certain equity allowance can be made (a problem which haunts the cooperative farming communities, like the famous Matador Farm in Saskatchewan, where the need to put something in trust for the children who may wish to leave always raises the equity or shares issue). Now, certainly the communal template emphasizes sharing of property, and, generally, minimizes personal property, but the precise arrangements within these boundaries can vary.

Another feature of the template is consumption austerity. A specialized version of this—asceticism—formed the basis of monasticism. An ascetic ideal induces sharing since the scarcity principle is involved, but this is a case where communalism becomes a kind of pragmatic outcome or artifact of something else. In the main line of communitarian movements, austerity is conceived as necessary to avoid distractions and prevent separation from God or other principles—and also to fight the capitalists who would enslave you. A characteristic pattern of commune evolution, now visible in the Hutterian colony and the kibbutz, is to permit *collective* consumption to grow, but to continue to enforce *personal* austerity—hence the kibbutz swimming pools, or the Hutterian air-conditioned grain combines.

In any case, probably a majority of communes had to exist in a relatively hostile and unsympathetic world, and therefore their resources were always relatively limited. Frugality is a necessity for people who must live on marginal resources and who may refuse help from the outside for fear of corruption. In any case, such austerity is always given a special justification, usually centering around the need for man to free himself of the distractions of want—that familiar Oriental idea. However, the fight against acquisitiveness can be waged in other ways than the communal—for example, by charity—and it is probably one of the few communitarian ideals that stand a chance of being taken over by many individuals in the corrupt world and made a matter of personal lifestyle. Still, it is hard, since it is much easier to live frugally when the people around you do also. Lurking somewhere in the background of the communal movement is a George H. Meadian [4] recognition that man is weak; he needs the recognition of others in order to gain identity. He is a sucker for conformity, and this works for, as well as against, the success of the commune.

Another paradoxical element involves the point of beginning of the commune. In most cases, the people who enter into the covenant were raised in a world of individualism and private property. I have noted that communalism sometimes just grows—a natural outcome of exclusion and scarcity—

but in the majority of cases, it is deliberate (hence the term "intentional community"). The idealists thus enter their putative communal world with habits that usually don't fit, and the initial tensions are always severe (the brouhaha in the first and only year of Robert Owen's New Harmony is a classic case, as Arthur Bestor has made clear [5]). Three generations probably constitute the minimum time period required to root out the old habits and inculcate the new; three generations are par for the course in nearly all social changes—another human rhythm rooted in the patterns of communication between parents and children and the difficulty of changing these patterns *in toto* in one generation.

Another source of paradox is leadership: the fact that so many of these groups espouse egalitarianism, but find that they need some degree of authority. The most common solution is charisma: the group either begins with a leader with strong, benevolent qualities, or it soon finds one.[6] Why? I believe that the tendency for communes, despite their egalitarianism, to so frequently resort to authority is based on two things: first, the need for a wise and persuasive figure to weld the disparate personalities of the commune's founders into an integrated and cooperative whole; and second, after the commune is an established fact, the need for a person with similar qualities to maintain the highly integrative and communicative set of rules of procedure. The commune is a very complicated form of social life, and one that does not always function automatically. Humans are refractory; they go off on tangents; they do not always follow the rules even when they believe in them, and they need someone to bring them into line. Hence, I consider that the commune, as an especially rule-ordered form of *gemeinschaft,* requires strong leadership, or at least finds it inconvenient to dispense with it, despite its frequent adherence to an ideology of egalitarian automatism.

The Rappites, a true but in many ways an aberrant commune, endured for a long time because of the special charisma of Father Rapp, who even took to building secret tunnels under the campus so he could emerge spookily in various buildings, impressing his "followers" with his mysterious powers; never, of course, clearly defined or admitted. And if one must look for a single explanation of the fiasco at New Harmony, it lies in the fact that the supercharismatic founder, Robert Owen, failed to appear during the first and only year of the commune's existence. Owen probably could have held the bunch together. The Hutterites, coming out of a late medieval tradition, thus lacking the extreme egalitarianism of the later secular movements, automatically provided for patriarchal leadership and benevolent charisma. This practice has served them well, although they have had to hedge it around with controls, lest it slip over the edge into petty tyranny.

Still another paradox centers on *kinship*. All communal societies, as we noted in the discussion of child rearing, find it necessary to limit the functions of the family and kin group. But at the same time kinship has played an important role in nearly all communal societies, as the primary model of the basic communitarian relationship: the egalitarian brotherhood. This particular relational form is sanctioned by special beliefs, but it also arises naturally out of activity-sharing, especially the work group. Women working together in a kitchen; men working together in the fields or in the shop—these are the archetypical communal experiences, and they all symbolize the band of equals helping each other in a brotherly way (or sisterly—we had better get the distaff side in here these days; unfortunately, the imagery of the old communitarian rhetoric is wholly patriarchal-masculine).

However work-oriented the concept of brotherhood may be, the cultural definition of egalitarian relations in most communal societies is based on blood kinship: from that kind of give-and-take loving-equality which is supposed to emerge in a family out of mutual support and affection. In more traditional groups, like the Hutterites, the band of male siblings —brothers—constitutes in fact the most important social unit in the commune, and this bond is extended symbolically to the entire group (although in the Hutterian case the women are excluded from participation in policy-making).

At the same time, the peculiarly complex ties common to kin constitute a latent threat to the commune. The symbolic extension of the brotherhood relationship is, well, symbolic, something that works fine if everyone believes in it, and especially if it is reinforced by some of the activity-sharing experiences mentioned before. Families easily become factions; brothers can stick together against other brothers. The most common factional disputes in Hutterian colonies are those between nuclear families, which, in the Hutterian case, are unusually large (eight to twelve children). The colony erects complex institutional controls over excessive kin-group solidarity, and this, then, is the paradox: kinship may be the basis of the commune, but it is also one of its threats. A choice must be made between emphases on two kinds of *gemeinschaft:* the family or the community.

III.

Margaret Mead insists that the contemporary generation gap is an unprecedented situation [7]: never in the history of human society has there existed such cultural divergence between parents and children, such anomic difference in views of the world and styles of living. Whether the situation is unprecedented or not I cannot say, but I think there is little doubt that the history of Western civilization can be written as a history of the progressive enlargement of the institutional world outside of the family and kin group, and it is quite possible that this trend has reached a point of crisis insofar as the young now can find a nonfamilial culture which challenges the culture of the family at every point.

I want to backtrack a bit to explain the significance of this for the communitarian movement. Some basic anthropology will help. As Earl Count has noted,[8] from an evolutionary standpoint the human nuclear family of parents and children develops out of the basic mammalian family and, while the mammalian family separates reproductive and nonreproductive activity phases, the human version does so to a greater extent than any other. Moreover, an additional human feature is the symbolic-cognitive representation of family roles:

these are not simply physiological and behavioral, but concerned with affection, hate, duty, responsibility, loyalty, property, and the mores. That is, piled on top of the biologically based role structure are a whole new set of often conflicting understandings and definitions of what it is to be a parent, a child, a sibling, a relative. The situation is such that the family or its enlarged kinship extensions can become society in microcosm, and for several millennia the kin group was, in fact, the mainstay of society; it contained within it most of the effective functions and roles. The portrayal of this mode of social life is the outstanding contribution of ethnology.

But there is another human feature: the tendency for "culture"—institutional life—to proliferate outside of the kinship unit. While family and kinship may have dominated for a long time, it is equally true that no *Homo sapiens* society ever completely lacked some institutional separateness in the spheres of "economics," "politics," "religion," and so on. It stands to reason that rivalries between the functioning kin group or nuclear family, and these adult associational patterns, would develop, and another major theme of ethnological research has been to show how this tension was managed in various human societies. In some, the solution was to permit kinship to move outward into institutions, or, more accurately, to construct institutional roles on the model of kinship roles. In the case of Japan, this resulted—in the post-feudal period of the late nineteenth century and still, to some extent, in a whole modern society—in the bureaucracies of government and business, work roles, the Emperor-subject dyad, and many other things being modeled on kinship roles, either quasi or actual. This Japanese situation seems to have been a rather unusual one, based on the very rapid development of an industrial nation out of a well-organized feudalism.

In the West the tendency, on the whole, has been in the other direction: to progressively narrow the family-kinship functions and permit role definitions in the secondary institutions to develop their own momentum and cultural character (this process is now under way in Japan also).

John W. Bennett

Hence, the tension between family and society, never resolved completely in any society, has perhaps been accentuated in the Euro-American system by unrestricted institutional growth. Problems arise particularly when the extrafamilial systems do not pay sufficient attention to the need to handle child socialization. If the extrafamilial world provides distractions and roles for the child before he is essentially humanized, there will be trouble, and that, in the minds of many analysts of child socialization, is the situation we confront.

Now, if this tendency toward extra-kin group cultural development has been a marked pattern in Western civilization from classical times to the present, we may have a part of the explanation for the high frequency of communitarian responses in the West, and their low frequency in the rest of the world, where kinship-based roles seem always to have had greater penetration into the society at large. The theory goes like this: the communitarian ideal is an ideological response to the need for continued socialization; the communal society is a concrete attempt to continue or resume the socialization of the human animal, established by people who despair of the ability of the family, in competition with the Outside, to manage alone. Out of this world then emerges all the kinship symbolism of the commune: the brotherly love, the sibling-brotherhood tie, the domestic sharing, the helping hand, the benevolence of the Patriarch, or the charisma of the founder or leader (remember that the major traditional Western kinship role is the Father). The commune would be, in this theory, an attempt not only to socialize the actual children in it, a task to which it devotes great care, but also the adult "children" and, especially (if it is a new commune), the founding generation of alienees from the Outside. These people know something went wrong out there; they feel they are victims as much as idealists; they are full of guilt at the failures of society. These reactions are clear enough in the contemporary experimental communes. In the established forms, like the Hutterian colony or the kibbutz, the guilt is usually buried and masked; perhaps even allayed, or sometimes transcended by the feeling that we have conquered, even though the rest of society is lost in the wilderness of suprafamilial structure.[9]

Victor Turner[10] has written something recently which seems to echo these interpretations (and those of Tönnies[11] and many others). He speaks of "communitas," a mode of behavior which rejects artificial structure, status, hierarchy, and seeks to go back to spontaneous holistic unity, the satisfying generalism of the oblong emotional blur of fellowship and brotherhood. He sees this search taking place in many societies, and I do believe he is right; that is, that this is part of the general condition of man. But I also believe that the West has had an especially bad case. Whatever the history, however, Turner appears to be saying that social structure is the extrafamilial system of rational status and role, largely devoid of the effect that develops in the *gemeinschaft*—family, kin group, village. I think he overdoes it, because there is plenty of ethnographic evidence to show that men can torture themselves and practice rational subdividing cruelty in tribes, kin groups, and villages, in spite of their communal institutions. However, there is something here nevertheless. The search for communitas is the search for the commune, or the search for the commune is the existential expression of the need for communitas.

If we review the slogans of the communitarians, we find that they specifically reject the Institutions. They are anti-Church, and seek to return to spontaneous worship; they reject Economics (which always introduces structure, status, acquisitiveness), and seek to return to self-help and self-subsistence; they reject Government and go back to the group-decision process of the family council; they reject Consumerism and the titillation of the mind by externally originating stimuli from the market place, the media. But why labor the point? I think that the significance of this cognitive map of 2,000 years' duration lies in the fact that there has been a disease in the socialization function. The commune tries to redress the balance between the family and the Institutions, to bring love back into the relations of humans by bringing back *gemeinschaft* functions and roles.

These themes emerge more clearly in some communes

than others. The Japanese commune mentioned earlier was a clear case: a small group of nuclear families was ostracized by their village; they banded together under the leadership of a benevolent and manipulative communitarian, and the commune soon became one big family, even supplying new conjugal functions in two cases of divorce in the original family units. The generally small size of most communes, and the frequent failure, as communities, of the larger ones, may be further evidence of the fact that these groups are as right as their own imagery; the life they seek is the life of the primary group, whether they are fully aware of it or not. Where larger demographic units have become the rule, and have endured, as in the case of the kibbutz, substantial change away from the primary group has taken place, and the commune comes to resemble a cooperative business more than the archetypical *gemeinschaft*. The Hutterites are somewhere between: the commune population remains around 125-150 at most (they divide at these numbers), but the special conditions of advanced and diversified production have also pushed them some distance from the original ideal of a familial brotherhood. This was the case in the sixteenth century as well—some Hutterian communes were as large as 400 and even 2,000 persons, in which case we are dealing with collective villages rather than the classic small commune. Hutterites considered these too large, and attributed some of their subsequent troubles to this factor. In the contemporary experimental communities, the desire to return to the *gemeinschaft* often fails, and in my view the failure often can be attributed to the failure to think seriously about primary roles: particularly the need for a male parental figure (or what we called earlier, in another context, a "charismatic" leader).

 Well, the attempt to translate the family level of integration into experimental living, in the midst of a big society that seems to go in the other direction, is not easy. Paradoxically, the positive valuation of the family must be countered by control of its functions. The history of every commune that has endured more than one human generation, and is thus in the business of socializing the young, is full of strife over this issue. The kin groups have persistent solidarity, and that worthwhile emotion, loyalty to one's own, gets mixed up with loyalty to the group, and the Institutions then begin proliferating *inside* the commune. Only time and endless patience can get the thing in balance. Again, I must cite the Hutterites as the polished case in point: the social management system of the commune-colony, at a population level of around 100, is more complex than the government of a typical small town of 5,000. Are we to feel sad that man cannot make it free of structure even in the most enduring and earnest circumstances? Or are we to congratulate the Hutterites for having clung to communalism no matter what, never mind the compromises?

I have a picture in mind: a windblown yard, with white-painted long houses in the offing, and bearded men, in little, widely separated clots here and there—one group up against the pump house, another in the machine shed, another just barely visible through the white-curtained windows of the chief minister's apartment—all talking, slowly, carefully, quietly, seeming not to do more than chat, pass the time of day as the brethren do as they go around their task circles, sometimes with little groups of women, even less conspicuously, doing the same. But no time-of-day this: what is going on are the caucuses of the nuclear families, working out their positions for the coming confrontation in the colony assembly, working out how they can make their point without an apparent and open break between these families, proselyting indirectly and gently, trying to realign the sides so the cleavage does not seem to pass down through the genealogy. This is not the ideal commune, but it is the human commune; conflict, when socialized, is not a threat—as Buber [12] perceives—it is the essential process of communal life.

Erich Kahler [13] and many others have stressed the loss of the sense of the *individual* in a world of giant organizations and structures; the problem, then, is not only loss of the socializing efforts on the incompletely gestated child by the family, but at a later age level, of the individuality of the human being. This loss of the sense of individuality creates great anxiety, fear, and a desperate attempt to do one's own thing: another insistent theme of the contemporary communes. We have, then, the two objectives of the contemporary commune: one, to restore the integrity of the small groups, the primary socializing units of society, in order to provide a full-spectrum culture with cradle-to-grave socialization in the same close-knit world for every individual; and second, to provide a sense of identity, a full and satisfying personality, for the individual members.

This second objective is, I believe, an essentially contemporary one. I can find no emphasis on individuality in the literature on the early movements with the exception of some of the utopian communitarian schemes of the late eighteenth and nineteenth centuries which also stressed equality over hierarchy (the older religious movement was more inclined to synthesize hierarchy with egalitarian brotherhood). Perhaps, in general, the individualizing motive is to be traced to this secular rational version of the communitarian template, with its intensive humanism, conscious philosophical exegesis on Man In and Against Society. The kibbutz is again a handy example: the movement began as a practical device to spread risks and to utilize more efficiently scarce social and financial capital, but with rapid infusions of socialist doctrine, it began to define the community as a place where the *individual* member could find peace and fulfillment. This led to an emphasis on *homes*, and to increasing comforts to attract people and keep them in the fold—most older kibbutzim now have their swimming pool, weekly movies, vacations, travel, and higher education possibilities.

These efforts, however, have not prevented a steady outflow of kibbutz personnel, and a chronic "opportunity cost" problem based on the fact that once the boundaries of gratification are crossed, the individual looks outside and sees something better ("but what have you done for me lately?"). The Hutterites stick to the point: they restrict consumption, specifically forbid the construction of *homes* (keep ornamentation, individual touches, hobbies, to the bare minimum), curtail movement, keep education at an eighth-grade level, and generally make it clear that communal living is communal and not individual. (They are greatly helped by the fact that they live in the northern Great Plains, where things are generally at a low level, hence cannot look around and see greener pastures.) Hence, Hutterites have a "defection" rate of not more than two percent.

I can hear some readers saying, "but he misses the point—we are not trying to establish a permanent community, only a place where for a while we can get away from the rat race and experience joys of *gemeinschaft*," etc. There is no doubt that most of the contemporary experimental communes will not endure longer than a generation at most, and that most of them, in the very words of some of their residents and founders, were not intended to endure. Consequently, the paradox may not mature to the point where one can speak of a *collapse*. That is, the "collapse" will be more of a withering, a simple decline foreseen by the founders, who felt that once the mission is accomplished, people can return to the Outside, purified and cleansed of the distortions. While I have no quarrel with these objectives, as a scholar of the commune I am inclined to say that the contemporary movement is an *application* of the principle, not a manifestation of it, resembling, in fact, the transitory medieval groups more closely than the serious experiments of the nineteenth century and certainly the Hutterites and kibbutznikim. While I am skeptical of how much of this partial communitarianism will remain, I also recognize the crucial need to experiment, to attempt an adaptation of the principle that may ameliorate the chaos of massness.[14]

Susan Rosenblum[15] sees the essential dilemma of the contemporary experimental communes in the conflict between "activist" and "privatist" roles. The activist wishes to *do* something—to use the communes as a platform for ecological, antiwar, prominority action—to realize the potential of the individual personality in constructivist opposition, which the commune is expected to symbolize. The privatist wants to fulfill himself by doing his own thing, by basking in the love and warmth of the primary group. Both are seeking personality and identity, but in different ways. The activist garners the accusation of betrayal of the commune; the privatist is accused of copping out of the good fight. Neither approach fully grasps the meaning of the communitarian tradition, since if that tradition means anything, it is that the commune stands (succeeds best) as an *example*, not as the blueprint of activist reconstruction, nor as a haven for the intellectually dispossessed, and in neither case as a place for personal motives of fulfillment. The commune is the *commune*; it is sharing, and, above all, it is the experiencing of complex and agonizing chains of adaptation that result from bringing intractable human beings into the milieu of share.

But it is *always* a matter of *bringing* them—not *making* them once and for all. This lesson, too, is hard to learn. The communitarian movement, like other products of cognitive idealism, is prone to the "substantive fallacy," or what the philosophers of science call "reductionism." It appears most clearly in the treatment of the concepts of "love," "work," and "share" in the movement, and it involves the assumption that if these things are manifested—if someone performs the act of love, shares something with somebody, or if someone works—a process of improvement automatically sets in. Love begets love; work begets integrity and clean minds; sharing begets altruism. The behavioral process is reduced to an automatic switchboard: push the love, labor, or share button and all else will follow, your problems are over. It is not that intelligent communitarians really *believe* this—they know human behavior is more complex. But a reconstructionistic idealism must deliberately seek out a simplified rationale or else it will founder in a swamp of uncertainty, anxiously chewing its nails as it contemplates intractable, unpredictable, perverse human nature.

And the substantive fallacy, like all fallacies concerning human behavior, has a grain of truth, if you consider the factor of *time*. If you labor, love, and share long enough, persistently enough, and in the milieu of unobtrusive leadership and unobtrusive rules of the game, you will get the idea; that is, "internalize" the habits. You may reserve to yourself those selfish motives and paranoid suspicions, never really rooting them out of the lower brain, but at least they are in the basement. You may need ways of seeking emotional relief—Hutterites take a lot of tranks—and the commune may come to have a slightly skewed, eccentric, moody atmosphere. Lots of colitis. Not only are the emotions being suppressed, but so are the guilt and doubt over the inevitable institutionalization of the system. I am reminded of one of the new communes that went through a crisis of divisiveness and wild accusation, and then came out of it with a strong leader and a new "constitution," full of arrangements for curbing paranoia and selfishness, mutual suspicion, refusal to take orders, etc. Things did not improve—the regulations were too obvious, too abruptly promulgated. The commune in its short life changed into a work camp, something closer to an old-fashioned Japanese rural construction gang, with their *oyabun* ("parent-figure") who demands loyalty and obedience in return for protection and a cut of the wages, guaranteeing every member a slot in a dormitory on the site, money if he gets sick, and so on. The line between patriarchal commune and the patrimonial lord-vassal household is sometimes extremely thin.

The enduring commune must be a matter of long participation, of continuous experiencing over at least three human generations, before something stable comes out of it. By "something" I mean, of course, a *changed* human being. First, there is the heterogeneous founding generation; they produce children who are the first to be raised in the commune; it is then *their* children, the third generation, who have the pure social inheritance of commune life, and it is this generation which has the New Men.

The New Men are those who have reached some kind of personal or interior settlement, restricted though this may be from the standpoint of the current hyperindividualist ideal; people who have also learned the discipline of the primary group: the hard daily tasks of survival and the willingness to do them without questioning; the ability to suppress suspicion and hatred in the interest of the collectivity; the ability to accept defeat philosophically, as the will of God or the will of the group. That is, *adapted* people.

John W. Bennett

In the adapted, three-plus generation commune, the eager and idealistic visitor from Outside looks for spontaneity, warmth, fraternity, enthusiasm, but he usually finds—well, perhaps a little bit of all of these, but also the need to avoid contact in order to reduce conflict, which can result in minimal or superficial communication; stringent rules controlling disputes and factionalism, which people observe in order to spare themselves embarrassment; the sheer familiarity of the commune group, people you know so well you need not converse with in order to communicate. Dullness is not the permanent state, of course; even the most adapted communes will have their recreation and enjoyment but, like most everything in a commune, sooner or later even enjoyment tends to become routinized—always, of course, by group consensus, never by fiat (not, at least, openly).

But I can hear the reader saying, "but what does he want? Is the commune a genuinely liberating, a genuinely socializing milieu? Can these 'adapted people' be considered people with an improved, A-1 socialized sense of identity?" I am afraid these are questions I really cannot answer; this whole paper is an attempt to lay out the complexities of the situation and invite the reader to try communalism himself, or to make his own study of the various forms. I *am* saying that communes can be a relief for people fed up, or unable to cope, with the rat race; and certainly, if the Hutterites are typical, true mental illness is rare in them. But I am also saying that to participate succesfully in a commune it is necessary to pay certain social-emotional costs (Hutterite psychosomatic illnesses are common). You have to make your own decision. One of these costs is an inevitable degree of disillusionment, a facing up to compromise and a bit of corruption of the ideals.

Now I have in mind another picture: this time it is a gentle slope in the deserty Southwest, a mesquite-dotted slope, with a dusty road winding toward the mesalike top, and into a cluster of rundown old ranch buildings, in various states of reconstruction or repair; and off to one side a new, amorphous structure of rough chunks of local stone, pine trunks, and plywood, smelling of linseed, big enough to hold the commune for meals and for meetings and recreations. In front of the ranch buildings are seated and lounging figures in colorful dress, each unique in his own way, the whole forming a style. Some are singing, others conversing, others playing with the children. A figure in Levis approaches and with quiet exasperation, very controlled directness, speaks to some of the frieze figures against the weathered old wood, asking for help on the irrigation ditch below in the valley, quietly reminding them that this is their scheduled hour. There ensues the usual short, bantering ambiguous argument, ending, however, with a movement toward the road. Thus emerges structure in the do-it-yourself commune, and if the commune lasts, so will routine and some disguised hierarchy and power.

I feel the need to say a little more about the contemporary movement and some of its major themes. The first of these concerns the family. While I have emphasized the preoccupation of communitarianism with the family and the need for socialization, the contemporary movement—at least some branches of it—singles out the family as the root of much evil, and advocates its demise. Following the early kibbutz, there is a belief that the nuclear family trains its members to achieve too hard, to pursue their lonely way without reference to others, to crave recognition instead of love, to expect victimization by capitalism and the mass media. In place of the family, the contemporary communalists wish to substitute a new group *gemeinschaft*—one that is sure to undergo some of the transformations and paradoxes we have described.

Linked to the family theme is the search for a freer expression of love and sex. These are the antitheses of the old ideal of monogamous marriage and the nuclear unit, since they deny that children need legitimation and that breeding can be uncontrolled without harm to the solidarity of the group. The pursuit of love, to make sex a free emotion, free of fear and prestige elements, is a worthwhile ideal—there is no doubt we have made a monster out of sex. But the communal attack on the family has deeper sociological significance: it aims to break up the straightjacket of the institutions—just as the sixteenth-century Anabaptists fought oppression and emerging capitalism by singling out the authoritarian relationship between ministers and their communicants as the key to the system.

Many of the contemporary communitarians also seek communion in drugs; in fact, the drug scene first emerged in the early 1960s in communal groups in California and on the East Coast. Why is it a key element in so many communes? Largely because of the attempts to break down the barriers between people that our alienating culture has erected—to smash these barriers, the communitarians seek instant communication, the group soul, and drugs certainly help. But in this, as in the sexual experiences, the commune cannot provide a permanent existence or even a model for general social reconstruction. The communitarians are not trying to remold the whole world; they seek only to make deliberate experiments with alternatives, hoping that out of the most extreme, as well as the more moderate, some new social forms will emerge.

Nothing, really, in the contemporary communitarian scene is new—only, perhaps, that emphasis on individual fulfillment I discussed earlier. The drugs appear in the medieval cryptocommunal and itinerant groups, in the form of narcotic mushrooms. The sexual experimentation, or something like it, is reported for the people of the Free Spirit, and for the Münster and other groups. The contemporary women's communes have their analogs in the Beguines. The nuclear family has always been a worry for the communitarians, and one, the Shakers, went so far as to eliminate it *and* sex altogether. While none of these extreme challenges to Western society have endured, they have all penetrated our consciousness.

IV.

Ralph Linton liked to remark, out of his deep pessimism, that no one has yet figured out a really good way to govern large populations, and that this is *the* problem of modern society, and of man in general. I doubt that the point needs arguing. If we look at the infrastructures of large modern nations, however, we may find that quite different patterns of social organization underlie the general structural similarities: the organization of the family and the relationship of the family to the secondary institutions; the techniques of social relationships in the society at large; the means of communication; and the criteria and allocative mechanisms of status. Among the variants appear to be some which remind us of the commune and its organizational strategies.

Chief among these is, of course, Japan. I have mentioned the tendency in Japanese society to use kinship roles as models for extrafamilial relationships and instrumentalities. To this can be added the fact that the fabric of Japanese society has been compared to a chain-link fence, with innumerable little close-knit circles of loyal brethren linked by obligation, and with these circles linked to others by similar obligatory ties. This kinship- and obligation-based system formed a nationwide network which permitted the Japanese to modernize, fight major wars, etc., with assurance of the complete loyalty and docility of the population.

Now, this kind of a social system has communal features. The family and its morality permeate the entire superstructure, translating the virtues of loyalty and the need to take the other person into maximal account, into social macrostructures. It is a society in which everyone knows his place; likewise, the commune. However, there is a crucial difference: the strongly hierarchal nature of Japanese society, made necessary by feudal traditions and the functional differentiation of tasks. Communal relations thus can be adapted to inequality and authoritarianism, and sometimes evolve toward hierarchy, authority, and functional differentiation. Thus, the Japanese case suggests that communal features can combine with totalitarian tendencies to produce a remarkably conforming and adapted population.

This suggestion finds an echo in the assertion by Norman Cohn that the millenarian movement is one of the roots of modern European totalitarianism. Cohn based his argument on the do-or-die mood of the communitarians, the prevalence of charisma, the fanatic righteousness, the use of primary loyalties and affectual ties that weld together the social fabric, the eschatological, timeless emphases in the ideology, the selfless ability to withstand persecution and criticism, the welcoming of martyrdom. The strategic use of these values and actions to organize mass movements in large populations is one definition of totalitarianism, although one could not possibly hold that totalitarianism is solely and entirely the result of applied communitarian models, and Cohn's thesis has been effectively criticized by E. J. Hobsbawm.[16]

The case of the Soviet Union must also be brought into the discussion. If communism is a secular version of the old historical communitarian tradition, and I believe that it is, then the communist state represents another attempt to foster cultural integration and personal identity with the use of communal techniques. The emphasis on the party, on myriads of groups and organizations devoted to reinforcing the doctrine and socializing the citizen and the young, the creation of communicative networks of authority, the controls on the family and kin group, and many other social phenomena remind us of the central themes in the communal tradition. But in the eyes of the true humanistic communitarians, the Soviet state is a monstrosity; the perfect manifestation of structure gone mad, out of control, suppressive of the true nature of man and love. From this viewpoint, no greater perversion of the communitarian spirit can be found, unless it is the Western mass society.

Contemporary Cuba also needs consideration. Castro's revolution contains a number of interesting and important communal features: the Isle of Pines collective youth farms; the Green Belt collective labor; the cooperatives of many kinds; the general spirit of equality and fraternity; the drive toward free necessities and even the extras. Certainly among the left-wing political experiments, the Cuban has manifested more of the old communal ideals out of the utopian tradition than any other, and this in itself is a desirable modification of the totalitarian matrix. However, the experiment is still new, it is supported by Russian money, and underneath the surface there is the hierarchy of the party elite, much uncertainty, privation, and compulsion. One must reserve judgment for a while, but the experiment is important and it, too, will become a precedent and a source of inspiration, even if it fails.

The noncommunist communitarian movements of today would appear to be about as far from totalitarian expressions as one can get. The emphasis on love, pacifism, and brotherhood, and the avoidance of structure and absolutism would appear to go in the opposite direction, and certainly no authoritarian political movement has yet emerged from the movement. But this is not really the point. What is meant by the Japanese and Soviet cases is that all human behavior is subject to transformation into end-seeking strategies; that is, it can become a *means,* and the dialectic of human behavior and morality is such as to permit the most desirable, humanistic tendencies to be utilized as means toward destructive and exploitative ends. It isn't that the commune is a danger; only that there are dangerous potentials in human behavior, or at least there are always some men who are ready to adapt anything to their purposes.

But there is another theme in the problem of the commune and the mass society. This concerns the search for personal identity and cultural integrity. Here we can make some profitable comparisons between Japan and the United States. Japan is entering her period of massness, consumerism, the pursuit of pleasure and individual achievement, out of a quasi-kinship and quasi-communal social base. That is, despite the many changes in Japanese society toward the open form, with its problems of incomplete socialization, youth rebellions, crime, and so on, the society probably remains more orderly and more amenable to control with minimal bureaucratic apparatus than the American. Is there an identity problem in Japan? There would seem to be, as the society leaves its communal past behind. The roots are still there, but they may be withering fast as "democracy" penetrates. The amazing spread of the "new religions," with their clear-cut rituals and group participation are interpreted by all students as substitutes for the old kinship and obligation circles, which are breaking up as the economy expands and encourages individualized careerism, success, hedonism. The symptoms of youthful delinquency and revolt are both evidence of the breakup of the kin group and its functions, and also evidence of the tendency for Japanese to move together into tightly knit bands. The student movements of Japan are quite different, much more solidaristic and "organized," than those of the United States. And true Western-type communes have begun to emerge. All this can be interpreted as the utilization of communal traditions to combat the breakup of communal organization.

In the United States, an approximately opposite process may be taking place. The society lost its folk-society-small town communalism long ago, and lacking a feudal tradition, never really had much of a communal emphasis—as de Tocqueville observed in the 1820s. But the continued movement in the direction of massness and individualism has clearly resulted in huge unpaid emotional debts: America may have the most acute problem of identity of all the large nations. Hence, the search for the communal milieu becomes a major movement—in the form of free-floating values and lifestyles, as well as in the more concrete phase of the commune. Thus, while Japan appears to be losing her communalism, America starts to seek it. And both societies are experiencing difficulties in holding on to it, and finding it, since the macrostructural processes are similar.

Where does this leave us? First, with the feeling that the communitarian tradition is something very human and deep-rooted—not simply the dreamy schemes of a handful of dropouts and children. It is, in fact, very this-worldly, since it points to the important need of *Homo sapiens* for the group as a source of experience and meaning. The commune is not a heaven on earth, it is the human on earth. As Buber is, I believe, trying to say, the social life of the commune is an especially intense, particularly typical version of the life of the micronetwork society; its government is the government of any small community; its politics are the ingrown strategies of the village or kin group; its economics are those of the co-op; its socialized "dullness" is the dullness of any dense population unit, never mind the gross size. These things all have their social costs as well as their identity-forming and culture-deepening qualities, and man's cognitive abilities will always require the payment of these costs.

However, men have repeatedly come to feel that the costs of the communal life are less than the costs of massness and alienation, and this conviction is once more prevalent and meaningful. Clearly, the world cannot be turned into one gigantic commune, but we could certainly use more *communes;* and the idea of .community, the adaptation of the communal tradition, has real possibilities, as thousands of people are discovering. We need more experimentation, more adaptation of this centuries-old critique of certain enduring problems and stresses in civilized life.

NOTES

[1] NORMAN COHN, *Pursuit of the Millennium* (New York: Essential Books, 1947, and other editions).

[2] GEORGE WILLIAMS, *The Radical Reformation* (Philadelphia: Westminster Press, 1962).

[3] EDWARD SAGARIN, *Odd Man In: Societies of Deviants in America* (Chicago: Quadrangle Books, 1970).

[4] GEORGE HERBERT MEAD, *Mind, Self, and Society* (Chicago: University of Chicago Press, 1934).

[5] ARTHUR E. BESTOR, *Backwoods Utopias: The Sectarian and Owenite Phases of Communitarianism in America* (Philadelphia: University of Pennsylvania Press, 1950).

[6] I have benefited from some pertinent suggestions from Professor Kurt Wolff on this leadership issue.

[7] MARGARET MEAD, *Culture and Commitment: A Study of the Generation Gap* (New York: Doubleday and Co. and Natural History Press, 1970).

[8] EARL COUNT, "The Biological Basis of Human Sociality" in *Culture: Man's Adaptive Dimension,* edited by M. F. ASHLEY-MONTAGU (New York: Oxford University Press, 1968).

John W. Bennett

⁹ Erik Erikson's hypotheses about the personality of hippies bear on the argument. Erikson proposes that the hippie-type personality features a high component of "basic trust," itself a part of what he calls the "first developmental position" acquired by the child during his period of "intense interaction with his mother (or, more generally, the early years in the small nuclear family). Basic trust "assures that *hope* becomes the fundamental quality of all growth." ("Reflection on the Dissent of Contemporary Youth," *International Journal of Psychoanalysis*, volume 51, pages 11-22.) In the rebellion against the state of the world, this becomes translated into an emphasis on love, fidelity to the good cause, and kindred virtues, and the young people who are supposedly most prone to this would then manifest dependency and the ability to suppress hostility toward others. These behaviors are, of course, important traits for the member of a commune. Combining this with the sociological approach already sketched, one could hypothesize that many individuals who seek to participate in contemporary communitarian movements, in or out of actual communes, are those whose socialization was incompletely developed beyond the "first developmental" level. Such curtailment would be the result of youth cultures, departure from the nest for college, or other mechanisms deriving from the secondary institutions beyond the family circle. The fact that the great majority of hippies come from middle-class homes suggests that the intense sheltering patterns and emphasis on parental love characteristic of the middle class have cultivated an unusually strong "first developmental position" in these people — an observation familiar to readers of child socialization literature. Persons socialized in lower-income groups do not experience this prolonged effect and, in addition, pass through much of their socialization in the streets, where the interaction tends to produce more open hostility and greater independence. However, the difficulty with psychological explanations of this kind is that they seem to label the individuals: in this case, as infantile. Surely, the communitarian movement as a whole cannot be explained in this manner, and the validity of love, brotherhood, and sharing can be established without resorting to interpretations based on patterns of emotional development. I therefore do not wish to push the argument.

¹⁰ VICTOR TURNER, *The Ritual Process* (Chicago: Aldine Publishing Co., 1969).

¹¹ FERDINAND TONNIES, *Fundamental Concepts of Sociology* (Gemein-

schaft and *Gesellschaft)*, translated by C. P. Loomis. (New York: American Book Company, 1940).

[12] MARTIN BUBER, *Paths in Utopia* (Boston: Beacon Press, 1958).

[13] ERICH KAHLER, "Culture and Evolution" in *Culture: Man's Adaptive Dimension*, edited by M. F. Ashley-Montagu (New York: Oxford University Press, 1968).

[14] This whole issue of the individual is, in fact, generating a basic division among the radical critics of American society: on the one hand, there are those who see rampant individualism as the source of all the evil (Philip Slater), and who want us to go back to the small group; and on the other, there are those who regret the loss of true individualism, the thinking, anarchic Independent Man (Richard Sennett), groups or no groups. Yet these positions are not as clearly separated as they might be, and the contemporary communitarian movement often tries to combine them. Clearly they have to be combined—a general society organized purely on the basis of the individual alone or the small group alone is unthinkable. The tilt in the direction of the individual produces the alienated chaos we are concerned about; the tilt in the autonomous group direction produces Japan, where strong authoritarian trends develop to organize the groups.

[15] SUSAN ROSENBLUM, "Designs for Survival or Revolution," unpublished paper presented to a seminar at the Center for the Study of Democratic Institutions, Santa Barbara, 1969.

[16] E. J. HOBSBAWM, *Primitive Rebels* (New York: Frederick A. Praeger, 1959).

REFERENCES

BARKIN, DAVID, and J. W. BENNETT. "Kibbutz and Colony: Collective Economies and the Outside World." *Comparative Studies in Society and History*, 1972.
BENNETT, JOHN W. *Hutterian Brethren: The Agricultural Economy and Social Organization of a Communal People*. Palo Alto, Calif.; Stanford University Press, 1967.
BENNETT, JOHN W., and IWAO ISHINO. *Paternalism in the Japanese Economy: Anthropological Studies of Oyabun-Kobun Patterns*. Minneapolis: University of Minnesota Press, 1962.

BLAU, PETER M. *Exchange and Power in Social Life.* New York: John Wiley and Sons, 1964.

BOGUSLAW, ROBERT. *The New Utopians: A Study of System Design and Social Change.* Englewood Cliffs, N.J.: Prentice-Hall, 1965.

COHEN, ERIC. "Progress and Communality: Value Dilemmas in the Collective Movement." *International Review of Community Development,* 1966.

DARIN-DRABKIN, M. *The Other Society.* London: Gollancz, 1962.

DELORIA, VINE, JR. *We Talk, You Listen.* New York: Macmillan Co., 1970.

ROSNER, MENACHEM. "Communitarian Experience, Self-Management Experience, and the Kibbutz." Mimeographed paper, n.d.

SENNETT, RICHARD. *The Uses of Disorder: Personal Identity and City Life.* New York: Alfred A. Knopf, 1970.

SLATER, PHILIP E. *The Pursuit of Loneliness: American Culture at the Breaking Point.* Boston: Beacon Press, 1970.

SPIRO, MELFORD E. *Kibbutz: Adventure in Utopia.* Cambridge, Mass.: Harvard University Press, 1956.

SUGIHARA, YOSHIE, and DAVID PLATH. *Sensei and His People: The Story of the Building of a Japanese Commune.* Berkeley and Los Angeles: University of California Press, 1969.

TALMON, YONINA. "Pursuit of the Millennium: The Relation between Religious and Social Change." *European Journal of Sociology:* Tome III, Numero 1, 1962.

——————. "The Family in a Revolutionary Movement: The Case of the Kibbutz." *Comparative Family Systems,* M. Nimkoff, editor. New York, 1965.

WOLFE, TOM. *The Pump House Gang.* New York: Farrar, Straus and Giroux, 1968.

Social Identity and the Formation of Social Movements

Alain Touraine

Laboratoire de Sociologie Industrielle, Paris

Sociologist. Specialist in social stratification and mobility and social aspects of economic development. Director of the School for Higher Studies, Sorbonne, since 1958. Author of L'évolution du travail ouvrier *(1955),* Sociologie de l'action *(1965),* La Conscience ouvrière *(1966),* Le Mouvement de mai ou le Communisme utopique *(1968),* La Société post-industrielle *(1969).*

Between Heaven and Hell

Sociology is the enemy of the ego. The consciousness that the social actor—individual or collective—has of himself certainly does not provide the meaning of his situation, the *raison d'être* of his conduct. Durkheim, in his study of suicide and in his methodological writings, showed that the study of representations should refrain from resorting to subjectivity in order to treat social acts like things. Marx, even more so, had performed this very reversal by criticizing categories of economic thought—that is, the consciousness of economic actors—by tearing off ideological masks and seeking the functional laws of an economic system. Still more impressive is Freud's demonstration. Who would dare, after him, to consider the ego as the principal organizer of experience, which makes the order of consciousness reign over the disorder of sensations?

Social identity is nothing more than the interiorization of values that are inseparable from the dominant ideology of a society. A manual worker, a young man or an old man, a member of an ethnic minority or of a dominated society may feel marginal and recognize that he occupies a low position in the scale of income, professional level, education, influence, etc.

What does this consciousness signify if not the recognition of a certain established order and the masking of the foundations of this order: bonds of domination, power, or exploitation. These power relations are transformed into scales of stratification through agencies of social control which give them the stamp of institutional rules. Methods of socialization complete the work of institutions: they teach one to adapt to the society, to realize one's place in it, and to understand the rules of the game, in order to use those rules in the best way possible. A certain theology of the Middle Ages, in teaching each person his duties and his functions in the social body, played a role very similar to the one often played by the social sciences today. These have become, to use the term of Jacques Lacan, orthopedic techniques. The principal aim of sociology should be, on the contrary, to criticize illusions of identity, recognizing first in social conduct the absence of identity and even of consciousness.

The more the actor defines himself by his practices and his social relations, the more he is assailed by the consciousness of being deprived and dependent, by the absence of communicaton, by an arbitrary power.

Identity is imposed from outside. It doesn't tell me who I am and the meaning of what I do, but what I should be and the type of behavior that is expected from me under threat of sanctions. Identity is nothing but submission, heteronomous and alienating, to a power.

In a slowly changing society, the actor defined himself by his belonging to collectivities, and by his social roles; at the same time this consciousness of identity was undermined by the constant reminder of basic contradictions (as expressed, for example, by the Christian ideas of the fall, sin, and redemption).

In a rapidly changing society like ours, where social ascription is increasingly losing its importance, in a society defined by its future rather than by its past, by its change rather than by its rules, social identity increasingly loses its content. Long before sociology, art exposed what Weber called the disenchantment of the modern world, relentlessly destroying the consciousness of the classical age, obliterating expression, refusing characters, as well as the illusion of the continuity of time and space. Art has become "abstract" and moved away from "humanism." The search for identity must be replaced by the will to control change. The aim of analysis can no longer be to understand the essence of man, but his forms of action.

But we must not consider the peculiar aspects of our society too hastily. Instead of beginning with an individual or collective ego, it is necessary to recognize a basic opposition: on the one hand, a society acts upon itself and creates its field of experience, beyond a simple reproduction of itself; on the other hand, it mobilizes resources whose structure cannot be changed.

Any society, from the moment it accumulates and invests (and is thus engaged in the work of its own transformation, hence possesses what I call a certain historicity), is at the same time drawn to certain orientations and submitted to "natural" constraints.

On the one hand, in effect, society formulates an image of its own creativity that depends on its capacity to transform itself. When this capacity is weak, creativity can only be grasped abstractly as an idea or "metasocial" principle, which is often called God or the Sovereign. When this capacity is strong, creativity is grasped concretely—it is called progress, science, development. But, on the other hand, society is also the implementation of resources—human and nonhuman, which possess, like every element in nature, their functional laws. In slowly changing societies, social and cultural organization is dominated by structures of exchange which present-day social anthropology allows us to discover. In modern societies, social organization is determined by models of development. But these societies are not freed from all obstacles in their promethean effort. They are limited by the internal laws of the ecosystem of which the social system is a part, and by the anthropological nature of man: by the structure of his unconscious life, of his sexuality and aggressivity.

Every society is, therefore, defined by the tension between

its self-transformation and the laws of "human nature." Every society is, like the tympanum of a cathedral, situated between God "in majesty" and the structure of hell.

Nowadays, as before, social consciousness is seeking to escape this dialectic of action, to fall back upon the unity and identity of an integrated experience.

On the one hand, historicity is considered as the pole of light, as what gives shape and movement to "raw materials," and the unconscious is conceived of as purely wild, irrational energy. The model of a theocratic society and the Reign of Reason has given way today to the image of a scientific society, each player carrying out limited and rational choices in order to make the most of his advantages and build the best of all possible worlds. The search for an identity is thus an open and complex strategy, a progressive adaptation to change. Liberal and optimistic in vision, the actor defines himself by his chances of progress, by the amelioration of his advantages, the diversification of his choices. It is a world without shadows, a house of glass, where calculations, games, and applications of the sciences to expansion, replace the reproduction of former systems of exchange.

On the other hand, we witness the explosion of a cultural protest that seeks to eliminate the opposite model of society by recreating primary communities, or by calling for an unrepressed revolutionary expression of the libido, or by returning to natural equilibriums.

Let us not identify these two images of society with particular actors, still less with opposing social classes. Their interdependence and their opposition are a fact of our entire society, defining its cultural field; they are themes that all the actors combine in one way or another in their actions. Thus, the actor who would identify himself entirely with one or the other would no longer be definable by his place in society; he would be a player or a dreamer, not a social actor; he would be nothing more than a social abstraction, beyond relations of production, politics, and social organization. Social identity can only be the product of the action that combines these two sides of social activity—historicity and "human nature."

Alain Touraine

The Conflict

We are, here, as far as possible from identity consciousness, from an image of man and a morality; as far from morality as is religion; as far from social adaptation as is psychoanalysis.

Instead of starting from the established order, from its values and norms, its definition of statuses and roles, we have started from the dialectic of historical action, its "openness" as opposed to the integration of social order.

This is not to give up the search for the ego, the consciousness of social identity, but on the contrary to understand its formation and its limits.

The formation of social identity is only possible if the social order no longer appears to the actor to be an impersonal system, but a product of men's actions, a projection of social relations, through which a society gives shape to the control of social practices by historicity. Actors do not live in a society integrated by values. The fact that savings and investments are accumulated out of the cycle of production and consumption signifies directly that it is not the whole society that realizes historicity, that carries out its own transformation, but that it is put to work—like the accumulation itself—by a ruling class. This ruling class leads society along its cultural orientation, but also identifies its particular interests to these orientations and controls the entire society to use social resources to its own advantage.

In our society, the ruling class, which runs the great organizations of production and development—public or private—manipulates not only production, but information and consumption as well.

This ruling class is not and cannot be directly conscious of its social identity. It is in its own eyes a servant both to public demand and to the possibilities offered by science. It is the instrument of the progress of a production and a consumption that simply respond to needs.

During the years when opposition movements were not formed or could not demonstrate, advanced industrial societies were dominated by the ideology of this impersonality of the ruling classes. Everything was analyzed in terms of social tendencies and instruments of progressive change: growth, modernization, differentiation, research, investments. The only problems of the affluent society seemed to be inflation and pockets of stagnation or resistance to change. The ruling class cannot acquire a consciousness of itself except in response to popular protest. We can observe today in the United States, in Japan, and in western Europe, that the economic-political rulers feel concerned by the urban crisis, pollution, the revolt of what they like to call minorities. The popular classes, on the contrary, may become conscious of their identity, not in terms of what they possess but of what they are deprived of, because they cannot identify themselves with a natural movement of society, from which they are separated by the domination that is exercised upon them by the ruling classes.

But those who do not control the economic and social development respond to domination not only by defending their social and cultural, individual and collective vested interests, but by claiming control of the process of change for the entire collectivity.

This constitutes a double effort to associate the struggle against bondage and the defense of present possessions. The dominated groups lean back on their traditions to foster their fight for liberation and progress. Rulers always say to those who are subjugated by them that they should modernize themselves, live with the times. If they did so they would lose their identity and be, in the precise sense which should be given to this term, alienated. Alienation results from a contradiction between the "dependent participation" through which an actor plays a role defined by the ruling class and corresponding to its own interests and the independent attitudes of the actor, who is interested both in defending his autonomy and in protesting against private control of collective resources. Identity consciousness is never the consciousness of the present; it is the invention of history, the mobilization of given resources, of what one too often likes to call the traditional culture, to regain control of the future. Nothing is more important than this dialectic of past and future, this zigzag that unites tradition and innovation through revolt, conflict, and hope. It is in defending his trade and his employment that the skilled worker engages in the struggle for the control of industrialization. It is, as Fanon and Berque have said, by leaning on the cultural and social strongholds that were the least contaminated by colonialism that national struggles for independence and development were formed.

This tension is so strong that the link between the past and the future, ascription and achievement, often breaks. Hence, the conflict between a reactionary defense of the "community" and the utopia of an entirely new and open society, between a closed past and an undetermined future. But each time this break occurs, each of the opposed orientations, which should permit an easy identity consciousness, leads to the dissolution of identity: the moderates integrate, blend together, and are no longer anything more than a stratum, a level. Radical isolation leads to schisms, internal struggles. Identity is not only born from the consciousness of contradictions but also from the search for control of social change. Working-class consciousness was supplied by the workers' movement: a struggle against capitalism, but inseparable from the defense of trade and employment, on the one hand, and the desire for economic and social progress on the other.

Social struggle does not curtail unity, but binds it together. Outside of the social struggle exist only alienation and its illusion of identity.

The search for identity is not a reflexive behavior, the discovery of social statutes and of assumed roles; it is the birth of a social movement.

New Social Conflicts

The essential link between identity and conflict—the genesis of identity through conflict—is not peculiar to our society, but to all societies where accumulation is significant, where social change results from internal causes, and where the opposition between a ruling class and popular classes is of central importance.

What are the particular forms that social conflicts take in our kind of society?

When the transformation of the society is slow, accumulation is limited, the cultural model is more abstract than concrete, and the ruling class controls "public life"; that is, symbolic systems and institutions, which are relatively limited and specific. Moreover, it establishes its domination upon a social organization which is not the implementation of the model of development but, on the contrary, a set of ascribed statuses. For example, a society is controlled by a political or religious order, while rural life, dominated by the landowners, is organized in a relatively autonomous way. Kinship systems, customs and myths, methods of socialization constitute a social and cultural community that is subordinated to the ruling class, but not determined by its decisions.

The classes are thus real groups, and society is divided into two parts by the dualism of work and leisure, of master and slave. This dualism, this separation between internal rules of local community and the realm in which the ruling class's power is exerted, fades progressively away. More and more, the ruling class controls all aspects of social life; at the same time, this control is more flexible.

From the period of liberal capitalism to the present, we have been witnessing the increasingly rapid disappearance of the autonomy of labor. Industrial sociology analyzed very well this passage from a labor system based on the dualism of labor and capital—the latter dominating the former but leaving it its professional autonomy—to a new, much more integrated system to which the concept of organization corresponds.

Capital has been more and more closely associated with "management" from the beginning of the Industrial Revolution, but especially in this century. As methods of "scientific" management succeeded each other more and more quickly, they first transformed the forms of manual jobs, then the organizations of workshops. Today they are transforming the system of communications, the methods of data processing, and, finally, the mechanisms of decision.

At the same time, factors of economic growth are getting ever more diversified, progressively touching all sectors of social life. The time has passed when they could be reduced to capital and labor. The crucial roles of research, technological progress, planning, and the management of large organizations bestow growing importance upon training and education; factor mobility; decision-making and coordination capability; attitudes regarding consumption or savings; and upon determinants of the relationship between savings and investment.

Social domination formerly exerted itself in a clear and imperative way on a limited part of social experience; so, too, education imposed restraints and prohibitions but consisted, even more, of the recognition of social structures and cultural expressions belonging to a particular community. Today social domination has become both more extensive and more diffuse. Similarly, education has become less constraining, yet aimed at modifying all kinds of behavior according to the objectives and forms of social change as they are defined mostly by the power elite. An affluent society is first of all a society capable of a generalized action of self-transformation. This is by no means a society of consumption, even though it consumes much more than former societies; on the contrary, it is a society of investment, since the part of production not used for consumption but for accumulation and investment is greater than in an agricultural or trade society.

The jump from societies of scarcity to societies of affluence brings with it a change in the domain of conflicts between classes. The role of property diminishes as the separation between the cycle of production-consumption and the mechanisms of accumulation weakens. It is the action on the environment, the power of orienting change on the whole of social life, that has become the central arena of class relations and social conflicts.

Social domination assumes three new forms in particular. In the first place, great organizations exert increasing pressure on their members to become integrated. Not because the hierarchy is more rigid and the relations of authority more brutal (the opposite is true), but because these organizations, which are complex systems of communication, must act not only upon the quantity of work provided but also on the attitudes regarding the enterprise and on social relations. Constraints must be internalized. One must "belong" to a company. Secondly, this domination spills out of the sphere of production and spreads to those of information and consumption, through mass media or the "agitprop." Finally, the increasingly growing role of States, of their power and their strategic possibilities, reinforces imperialism, the will of world centers of power to dominate regions that are underdeveloped or incorporated into a sphere of influence.

These three themes are present in all great social movements of today, which no longer define themselves by an economic conflict, but rather by their opposition to cultural, political, and social domination.

When this domination is centralized and directed by a political organization, it can become totalitarian. But whether it is or not, the degree of social mobilization, the field of action of society upon itself, does not stop moving forward. Already, very clearly, quantitative change is transformed into qualitative, which allows us to consider our society, no matter what name we call it by, as different by nature from the societies that are still involved in the process of their basic industrialization.

Since the formation of the "black country" and of a large urban proletariat in England in the nineteenth century, the capacity of society to act upon itself has not ceased to grow. The field of social struggle has both expanded and been transformed.

Protest can no longer be sufficiently defined by the struggle against unemployment, low salaries, the irrationality of economic crisis, or the transmission of social inequalities.

In the most economically advanced societies, protest is mainly aimed against a system of social organization, both against the accumulation of power by organizations and

Alain Touraine

against the increasing manipulation of all sectors of social activity.

Whether the demands are only defensive or at the same time offensive, they attack a type of decision, the behavior of social actors, rather than "economic laws." Social struggles are more political than economic and attack a mode of development directly. This change is already felt in the case of the workers' movement, which maintains the form of a social movement but seeks to transform itself, as in Italy or France. This theme of self-management was the center of the strikes of May/June 1968 in France in the most modern sectors of economic activity. Still more evident, in the universities students rejected a system of authority that confronted them with knowledge as an established order, instead of considering knowledge as a creative activity.

Another essential aspect of these transformations, closely linked to the preceding, is the reversal of relations between the majority and the minority. The ruling classes have always been restricted elites, controlling and limiting their recruitment and opposing the masses, so that the largest popular movements have always cried out—against the elites and privileged—to the people, the nation, the majority. Socialist movements have constantly denounced the fifty or two hundred families who have run this or that country. The apologies of Mandeville or of Saint-Simon expressed this appeal to the workers against the ruling minority of the idle and the profitmakers. Today what must be added in extension of Galbraith's analyses is that the organizations themselves, and not just a small ruling elite, constitute the ruling force. What has been said of the increasing capacity that our society has to act upon itself brings with it as a direct consequence the integration of an increasingly larger portion of the population—both in the sector of production and in that of consumption—into the social organization, as it is designed by the ruling groups. A growing part of the personnel of large enterprises, as well as numbers of educators or simply consumers, participate in the system of economic growth and social transformation. If the phrase "the silent majority" is so successful, it is because it corresponds very well to the social situation of countries as different as the United States, the Soviet Union, or France. The old democratic appeal to the mass or to the majority is no longer anything but a conservative appeal to all those who participate in a dependent manner in the affluent society.

Let us sum up the previous remarks: before recent technical and economic transformations of the most developed societies, the forces of opposition defined themselves both as a majority and as of primarily one status (wage earners working for the capitalists or subjects of the Prince). Today they show themselves to be a group of minorities, struggling against the grip of domination in the most varied sectors of the society. This heterogeneity is all the greater in that the forces of opposition are resisting directed change, based upon their ascriptive attributes, in particular those not even directly linked to the role of production: age, sex, race, religion, etc.

It is true that this dispersion is compensated for by the increasingly extensive character of social domination, so that the actions of these minorities can converge against the system of domination and, on the other hand, be much more free from social and cultural particularisms than the actions of former popular groups.

Nevertheless, social struggles and, consequently, the consciousness of social identity appear today both more total and more dispersed, more fragmentary, than before.

But it is false to assert that today's social movements mobilize unorganized masses, reacting in an effective way, sensitive, above all, to the appeal of charismatic leaders. Mobilization is only important when it stirs real groups into action, groups whose members are linked to one another by a common social experience, such as the campuses, the ghettos, and large factories. These are centers for uprisings, hotbeds for class consciousness. Movements are not the result of the disorganization caused by increasingly rapid changes, but the expression of conflicts in which groups rise up against a concrete experience of domination.

Each opposition group tends, especially at the beginning of its struggle, to react to domination and to the grip of mass

Alain Touraine

society by "barricading" itself through the affirmation of its own unique qualities. But, as we have already said, such an attitude only leads to self-destruction, since social evolution has already largely destroyed the autonomy of social and cultural particularisms.

In a situation of rupture, one sees the minorities divided between two complementary and opposed orientations; on the one hand, an affirmation of the collective being, and on the other, an aggressivity towards the dominant order. In the same way, in the beginning of the labor movement, communitarian or cooperativist movements developed from Owenists and St. Simonians, while at the same time, violent actions against the social order were launched by Blanqui and Barbès, amplified by popular uprisings and revolutionary strikes.

The same duality of action was observed in the sixties in the universities: on one side, the violent action of Zengakuren, the movement of March 22, or of the SDS; on the other, the more communal activities of groups of young people. The two sides of the action are always more or less linked; a true social movement is only formed when this bond is strong. But an opposed, self-destructive combination can also happen: students' actions against the university, or sabotage by revolutionary unionists. These are signs of the failure of the movement, which is reduced to attacks against alienated behavior and to dependent forms of participation.

Protest behavior is never the direct affirmation of an identity. It is torn between defensive and counteroffensive attitudes. It is always divided between an alienated mass and a counterelite. Analyses of the workers' movement in its classical period have shown this very well. On one hand, an elite existed, that of skilled workers who were defending their trade, employment, and tradition, but who also played quite a central role in the system of production, and defended the cultural orientation of their society, the idea of progress against the capitalist leaders. On the other hand, there were the underprivileged, who underwent proletarization and would only act to protect their earnings, and would break with a system over which they had no control.

The workers' aristocracy was both the core of the workers' movement and the basis of reformism. The underprivileged gave the working movement its momentum, but did not play a determining role in its organization and political and ideological development.

The workers' identity is thus not a simple basic principle; it is only built by the bond and the tension of the two sides of the workers' movement.

The historic reminder allows us to approach the question which is asked today: where and how can such a link between a counterelite and the underprivileged appear; where can a social movement be formed from which the consciousness, always torn or tense, of a social identity be created?

Observation seems to show that the university is at least one of the privileged places where such a social movement can appear.

Because science and technology are today important forces of production, the university is no longer on the fringe of economic organization and political choices. These themes did not first appear in declarations of student movements, but in the *Bulletin of the Atomic Scientist* and, more generally, among circles of scientists who could not pretend to ignore the political, economic, and military effects of their discoveries. Sensitivity to this theme was greater in Japan than elsewhere; that country has the largest proportion of graduates who have entered business. It is widespread in departments of social sciences everywhere, since they, whose scientific rigor is still relatively weak, are much more linked to management or business than before, and often serve as instruments of social domination and ideological domination.

These dissident elites appear not only in the universities, but more and more in large organizations of production, where there are great numbers of professionals and experts who possess more competence than authority, and who thus dispose of a professional autonomy that distinguishes them from the executives and bureaucrats defined more and more by their places in a hierarchy of authority.

It is this dissident elite that is promoting the new social movement. But it is also fragile, like the former workers'

Alain Touraine

aristocracy; subject either to corporative defense or to moralizing illusions; or, more simply, to the pressures of success and personal careers.

But where are the underprivileged in the university? It is evident that the students are not, in the large majority, underprivileged, and that it is even difficult, given their very transitory status and their economic dependence, to place them on an income scale or on any other scale of stratification.

In certain countries, however, as in France, the gap between a rapid growth of the number of students and a very slow transformation of the organization of studies has created a serious employment problem. Many students receive degrees that correspond only to the needs of the school system itself (but far beyond the need to renew the corps of teachers), and do not correspond to other types of available jobs.

If we consider youth in general, it would be paradoxical to see it as a category of the underprivileged. A society in rapid change, on the contrary, values youth at the expense of experience. Surely the old are more underprivileged in our society than the young.

These few remarks suggest that there exist in the university insufficient conditions for the birth of a complete social movement. The protest elite that is found there must leave the university in order to meet the underprivileged categories capable of giving to a social movement a strong and active sensitivity to the contradictions of society.

In less industrialized countries, like Italy or France, students call first of all on the working class, which, in spite of the progress of collective agreements and social laws, is still submitted in the factories to a very autocratic domination and whose labor unions are still faithful to their revolutionary orientations.

Actually, it is the conjunction of this revolutionary attitude, fostered by the students' initiatives and by union pressure to push back managerial power and arrive at true negotiations, that explains the passage in these countries from student agitation to vast strike movements like those of

May/June in France or the hot autumn in Italy. In the latter country, a large proportion of the working population is of recent rural origins and thus has retained the revolutionary spirit that it had already shown in Turin at the time of the *ordine nuovo* just after the First World War.

The workers' movement is not in itself revolutionary at present. But there exist fervent revolutionaries in the working class: a rapid social change is actually associated at the same time with the domination of large firms that are making Turin almost a company town, where archaic forms of political and cultural domination persist. The current flow of economic change is pounding at the barriers of the social order, reinforced by the role of the Church in the case of Italy, and by the role of the State in the case of France.

In both cases, it is the young workers who participate most actively in social struggles, partly because they are more modernized, and at the same time because they are sensitive to the professional barriers against which they are pounding.

In the United States, these professional barriers are not as high and the hold of the State or the Church on social life is not as great; however, in this country the movements of the underprivileged are larger, more violent, more independent than in European countries.

This is so because if European societies rest on the strength of a political or ideological domination, America is dominated by values, by a morality built on the good conscience of pioneers, puritans, and local leaders. Mass society having come to the United States at its highest degree of development, the contradiction is all the stronger there between the moral and social values of "good citizens" and the push of the "barbarians," torn from their traditional collective identity, dragged along in the change, and rejected at the same time by the elite. These underprivileged, the blacks in particular, do not act like a people or a nation but rather like those whose identity rests in nothing more than a refusal to integrate, though offered integration by the dominant social group. They affirm their collective identity by a rupture with an underprivileged status.

They are the proletariat, not in a directly economic sense

Photo by Ken Heyman

but because they are both attracted and rejected, used and unrecognized.

In all cases, the main problem for the formation of a social movement is the meeting of the protest elite and the proletariat. Almost everything opposes these two categories, whose interests in a given social system are often opposed. Their meeting can thus not be made except through certain pressures: economic, political, or military crises.

In our society, drawn by economic growth and social change for twenty years, such unifying forces of a social movement could hardly be found in the most favored countries, although part of the European working class still lives under economic and social hardships. It was the Vietnam War, the blow given to American victoriousness by a people and by a party, that brought into being a series of critical reactions against the dominating system. Little by little this war destroyed conformist identification with a society and with its values. American society and, in reaction, the analogous societies of Western Europe found themselves confronted by an image of themselves that came from outside and contradicted their national ideology.

In just a few years, the image that the society had of itself appeared naive or a hoax. Every social movement supposes that it is not going to play the game anymore, denounces it as trickery, refuses to define itself by the place that it occupies in the system, and places itself outside of this system and against it.

In order for an individual or a group to become an actor in the transformation of his society, he must first cease to accept the identity that the social system gives him. He participates in a collective action, and acquires a new identity, by rejecting his statuses and roles.

Political Consciousness

Involvement in social struggles creates a social identity consciousness. But this consciousness is neither simple nor integrated. Each of the adversaries acts in several directions at the same time; a ruling class is both innovative and reactionary; a popular class is both conservative and progressive. It is also dominated even in its consciousness; alienated, that is to say, reduced to a dependent participation, as the ruling class defines it. The dominating class, on its side, flees the consciousness of itself by identifying itself with progress, rationality, and the general interest.

Social conflict takes on a greater visibility, the position and the nature of the opponents become easily recognized when they identify themselves in terms of a political struggle. You must vote white or red; you must choose one party or the other; you must support this or that side in a trial.

It is not on the level of class relations that a simple social identification can appear: dialectical social relations are expressed by a divided consciousness. A "positive" consciousness can manifest itself only when the actor is defined by reference to a principle of unity, to the State.

It is truly the concrete struggle for political power that constitutes the identity of the historical actors. It is clear that participation in political life can dissolve a collective social identity; the more so when political parties are not oriented toward extreme ideological attitudes and are complex coalitions made to win elections, internally diversified and little cohesive. Few countries had attained a higher degree of integration of social forces into the system of political representation than Uruguay: he who was not blanco was colorado, and very rare were the associations, clubs, or unions not immediately identifiable with one or the other side. In the same way, each party became so composite that the electoral law recognized it officially and invited the electors to vote for this or that "lemme" (faction), those dividing up afterward the seats won by the whole party. The opposition between the two parties, which was at first that of the rural society and the urban society became complicated to the point that political identity no longer had clear content and did not allow predictions of the social conduct of the members of the electors of this or that party.

Must we then see in the opposite situation, that of the nonparticipation in political institutions, that of revolutionary movements which are against the rules of the game, the condition favorable to the crystallization of political and social identity? Yes, in a sense. It is surely clearer to be a tupamaro than a colorado, to go back to the Uruguayan example, since the break with the institutional system provokes, by repression and encirclement, an identity consciousness.

On the one hand, however, this acute identity consciousness only reaches a small number, even if it has a great capacity for diffusion and influence; on the other hand, and above all, anti-institutional political action, withdrawn in the affirmation of itself, is subject to pressures of isolation and repression, to the difficulty of deriving a strategy from principles, to a constant tension between the purity of the ends and the efficiency of the means. It is constantly weakened by struggles between rival factions or organizations, by doctrinal heresies and organizational schisms, to the point that if the consciousness of belonging to a group is very vivid, the contents of the belonging are far from clear, because rival groups claim the same principles and similar objectives. The reference to the political organization itself can be more central than any other social status—as in the case of the bolsheviks.

Political affiliation doesn't create a social identity, because it means at the same time consciousness of belonging to a social group and participation as a political force that is defined by its role in a representative system.

This is why a social movement usually passes, in its political existence, through three successive stages: that of the anti-institutional break, that of the political confrontation, and that of institutional influence. The keenest consciousness of political identity is found in the central phase, in which a social force acts both outside and inside a political system, as was the case at the time of the great development of the workers' parties in Western Europe, in the first half of the twentieth century.

At present, are we witnessing in industrialized societies such a confrontation and political consciousness? It seems, on the contrary, that the social movements we can observe are either in their early or in their late stages of existence, none of them being at their zenith.

The workers' movement in most advanced industrial societies no longer creates a political identity. Political parties are not homogeneous social forms, and the distance between unionism and actual political action today is considerable.

As I have already indicated, it is only in France and Italy that the workers' movement has maintained any political importance.

On the other hand, a rupture has appeared between a sysstem of political decisions and the anti-institutional push of new social movements. Throughout the Western world, opposition is extraparliamentary. It is thus both utopian and violent. Utopian, for, not having any hold on the institutions and the political decisions, it can only oppose the present state of society with a possible or desirable state, proposed as a complete model of society. This new society cannot be reached through institutional channels but only after a complete breakdown of present societies. These utopias can be called neocommunitarian. They are not workers' communities, like those dreamed of or founded in the nineteenth century; they are nearer Fourier's ideas and his psychosociological utopias.

These "natural" communities seem to be entirely opposed to militant communities, defined by opposition to the outside world and the mobilization of every energy to defend a collective independence that is national or quasi-national. But these two types are not actually opposed to each other; one passes easily from Rousseau to Saint-Just.

A social movement being born in a state of extra-institutional opposition cannot enclose itself into the creation of a utopia, whatever kind it is. The utopians must look for alliances. The workers' movement in its beginning in France was both animated by sectors of utopian socialism and drawn into the republican movement. Even at the end of the nineteenth century, a comparable duality can be found between anarcho-syndicalism on one hand and "possibilism" in all its variants, which incorporates the social movement into political forces that seem favorable, at least liberal, even though foreign to the specific objectives of the workers' movement.

In the same way, militant minorities are sometimes ready to accept a "front," an alliance with progressive political forces, such as was seen in the United States at the time of the campaigns of Eugene McCarthy and Robert Kennedy, or even during the campaign for a moratorium in the Vietnam War.

The relations between these different forms of action depend first of all on the situation of the political system. If it is rigid, if problems and actors represent a former state of the society, the distance between the utopia and the "fronts" may be very great. This distance is all the greater when the political choices offered appear more limited and less pertinent. Political system and fields of social struggle exist today in many countries, certainly in communist countries whose political systems are highly rigid. If it is more open, as it is in many western countries, a social movement can have a more direct political representation. But if it is still loosely organized it can be marginalized, out of the broad limits of the political system. The more liberal a political system is, the more likely social movements are to resort to violence.

Between the workers' movement, which has turned into an institutional force, and the new movements, which have not yet arrived at a unifying class consciousness, the political system is empty, both active and almost unrelated to social movements. It is dominated, on the one hand, by the defense of traditional, local interests; and, on the other, by the role of a centralized economic and military power. In many countries it is both above and beyond the field of social conflict.

Social Identity and Organizational Problems

We have indicated that social identity can only be born from participation in the conflicts which form around the control of the processes of social change. But at this level each of the conflicting social classes is torn between opposed orientations; they defend some vested interests, inherited from the past, and at the same time create an image of a new society. It is the reference to the political system, to concrete decisions affecting the collectivity, that crystallizes this consciousness of social identity. But the latter is still not a simple affirmation or a self-definition; it is still divided between complementary but opposed behavior patterns.

It is now necessary to consider the problems of social identity and social movements at the level of the institutions themselves, in the city, factory, school, hospital, office. Here, where authority and sanctions draw a clear-cut line between what is accepted or rejected, between in and out, social actors are bound to define themselves in the same way, by accepting or rejecting a set of rules and regulations.

An organization mobilizes social resources; it is an agent of social change. It does not survive and develop unless it brings technical progress, a better division of labor, a more efficient system of communication, etc. This is its practical, technical side. But it also embodies a social power, with its values, and its double action of integration and repression.

These two aspects of an organization are interdependent but also opposed. For example, a university has a practical activity, since it creates and transmits new knowledge. In this sense, it is open, flexible, competitive. But at the same time, it provides an education, socializes the students to a concept of the culture which is linked to the structure of power in society.

It can happen that these two aspects are closely linked, in particular when the ruling class is rising. It can also happen that the practical function overshadows the ideological function, which is the case in organizations that are subjected to strong technological or economic constraints.

In others, the opposite case is true: the ideological function overshadows the practical function and manifests the hold of groups more oriented toward social integration than toward change.

Forms of opposition are very different, according to the manner in which the organization fulfills its practical and ideological functions. Let us outline these differences in the following schema, in which + signifies that a function is being fulfilled in an organization, and − that it is not. The forms of opposition are shown in the cells of the following table:

		Practical Function	
		+	−
Ideological Function	+	Protest	Rejection
	−	Utilitarianism	Decomposition

Opposition to an organization that fulfills both a practical and an ideological function leads to protest, to the affirmation of the actor as a participant in social change, confronted with a system of social and cultural control. It is in this situation that "positive" social movements are formed. If, on the other hand, the practical function fails, partially or totally, the actor no longer expects the organization to ensure him social participation. The university does not create knowledge or teach techniques; industry does not guarantee full employment; urbanization goes ahead of industrialization. It is then that reactions of rejection are formed, directly aimed against the system of dominating values, which is accused of inefficiency.

Economic activity no longer seems to be anything more than a concentration of wealth or speculation; the university is denounced as an ivory tower conceived for the intellectual comfort of a professional elite. In a parallel way, commitment to both the ideological and instrumental role of an organization results in an identification with the ruling class. The force of a social movement must then rest on its negativity.

This opposition of two kinds of collective behavior is translated into their forms of organization. In the first case, it is a question of a mass movement, supported by the spontaneity of its members, who are united by a common participation in a sector of activity and by demands. In the second case, the marginality or exclusion leads to the dis-

Alain Touraine

persal of members, who begin competing against one another to take advantage of insufficient opportunities offered by the system. The latter movement depends on the action of strongly constituted avant-gardes of the bolshevik type.

The two other cases can be disposed of more quickly. The combination of a strong practical function and a weak ideological function corresponds to an "open" society, which is strongly competitive, even brutal, but in which ideological crystallizations play a weak role. Society appears as a market where everyone seeks rewards for pragmatic efforts.

Finally, the case of decomposition corresponds to the situation of those who are not integrated into instrumental activities in a society, where consensus is low, power weak, and where the ruling class is not integrated. This is the case described by the concept of marginality.

No national situation, even in a particular sector of activity, corresponds entirely to any of these cases. But it is the study of combinations among them that explains their particular forms.

In France in the university, the phenomena of decomposition, rejection, and protest can be observed at the same time. The May 1968 movement marked the grip of protest upon rejection and decomposition. But very quickly, in the fall of that year, protest became disorganized, which gave more importance to the movements of rejection animated by ideological sects, relatively isolated in a sector in complete decomposition.

The American universities, on the contrary, experienced a durable protest movement, which freed itself from utilitarianism sometimes associated with conducts of crisis, but with some importance left to reactions of rejection, given the solidity of the academic system and its practical dynamism. On the other hand, American society is experiencing strong reactions of rejection in other sectors. For it is both a mass society where the degree of mobilization, in the way Karl Deutsch defines the term, is higher than elsewhere, and a society that is founded on the values of integrated communities and, through segregation, rejects out-groups. In particular, the penetration of blacks into the industrial and urban economy of the North is contradictory to the values of the ruling white society. Unemployment, the exclusion of the ruling professions, and residential segregation make this situation explosive.

Thus, passing from the first level of analysis, that of a general type of society, defined by its cultural orientations and its class relations, to the observation of concrete aspects of social organization, we proceed from complex relations between conflicting classes to the confrontation of "positive" social forces that accept and reject, and that must be accepted or rejected. Identity cannot be reached through social integration, except by conscious agents of the ruling class.

For the rest of the people, the way out of alienation supposes a combination of rejection of the power system and of a claim to control social change.

But it is natural, since the dominant order only *talks* of integration, participation, and collective progress, that the quest for identity be directly linked with withdrawal from the organizational rules and ideologies, with rejection of values and interests, and with denunciation of exploitation and alienation. If it is true that it does not stop there and cannot exist without incorporating "positive" demands, these can never be isolated from the keen consciousness of nonidentity. In the heart of affluent societies it is through the cry of the absence of identity that both the dominant ideology and the false identity of the silent majority are shattered.

The break will only lead slowly to confrontation, but it prepares it, as confrontation in its turn leads either to mutation or to negotiation.

Identity and Change

The preceding remarks can contribute to answering the question: what is the role of youth? In a society defined by its change and future rather than by its heritage, the young are at the center of social debate. While the aged are isolated and most often forgotten, frequently reduced to poverty, and are more of a problem than social actors, the youth are privileged and make demands. It is from this duality of situation that one must begin. This transitory status could be the most favorable today to the formation of a collective identity.

It does not suffice to say that youth is in an anomic situation, detached from the family and from the office, living for a longer and longer period of time in the academic system that puts off the moment when adult roles should be assumed. Such an analysis is not devoid of utility; it can explain the loss of identity, the crisis of socialization. But it cannot, by its very nature, explain the birth of collective orientations. It indicates where the nonsense lies, not where the new meaning of situation and behavior is.

Actually, if we can still talk of a youth movement, it is because youth, more sensitive to social change, has mixed in and sometimes mutually reinforced several types of reactions.

First, there is the *innovation*, the rejection of old ways of thinking and feeling, the attraction to new languages of consumption or communication. It is especially in favored youth, among students, protected from poverty and even from the constraints of early work, that these innovations develop. This group invents behaviors and cultural expressions that are an integrating part of the cultural field of the whole society. It opposes what Kahn now calls a "sensate" culture to the stoicism of the old generations, who were oriented toward work, savings, and rules. *Then*, there is *protest*, because students are more and more often future professionals who will have to work for large organizations. Some of them oppose the appropriation of knowledge by economic and political powers.

Finally, there is the crisis of young people who experience the gap between socialization agencies and social environment, and who seek desperately to preserve or recover personal identity in a world in movement.

This general dissatisfaction of youth is less marked in countries where authority is embodied in the State or the Church, in the bureaucracy or the school, than in those where it is more "moral," where the appeals to cultural values and to responsible personal choices are more constant.

This complex mixture of different types of behavior cannot but be unstable. Little by little the mixed elements separate. Social criticism becomes a struggle which imposes organized action and makes the weight of repression felt; cultural innovation achieves enough success to be easily adopted by the consumer society. The paths of retreat or evasion lead to isolation or deviance.

We have come, it seems to me, to the moment when the diversity of the first period may be either lost or transformed, when each element participating in the agitation of youth risks losing its strength by isolating itself. Innovation easily becomes a way of life of a new aristocracy or a vast café society; social revolt can be reduced to violence; crisis can be lost in the withdrawal.

Conversely, these elements can be combined; self-defense can be united with the struggle against the adversary and with the image of a new society. Then a social movement is formed. It is not necessarily integrated, conscious and organized, but it is capable of becoming an agent of conflict and social change. But such a movement can only appear by surpassing the limits of youth. It is formed inasmuch as cultural escape and social revolt become linked to one another, and by their union go beyond themselves.

Cultural withdrawal leads to a defense of the community, which challenges the political system responsible for the norms; norms that are rejected. Thus, in the United States, appeals to community become the strategic line of defense of the environment, shaping the conflict into one between big business and grass roots democracy.

On the other side, social revolt attacks the most general political decisions, imperialism above all, and from there descends to a criticism of particular institutions and to cultural mobilization.

This double movement reminds us of the process of formation of the workers' movement. It started both from the denunciation of crises or of unemployment and from the defense of crafts and of professional community. But today the field of social conflicts has been transformed. It can no longer be defined only by the labor situation, strictly by roles of production. It has expanded, as our society rapidly develops its capacity of action on itself. That even leads to a complete reversal of the battlefront. The proletariat used to protest in the name of labor against the leisure of the ruling classes, or against the heritage of social inequalities. Today it is the ruling class that speaks in the name of growth, of technical and economic rationality. And opposition movements appeal first to the defense of the human being, against the production system, to ascription against achievement.

It may seem difficult today, in the face of this disjointed picture of social struggles, to find their general meaning. But the time is still not long past when, confronting the shocks of Berkeley, Tokyo, Berlin, and Nanterre, those who appealed to the constant necessity of examining the conflicts and social movements seemed to be talking in a void. It is as impossible today to see in the new social problems the simple extension of those of the preceding period as it is to accept the idea that necessary social changes are accompanied merely by tensions, lags, and crises. These changes are not socially and politically neutral. They are the stakes of future social struggles that are still in their infancy today, not having attained the degree of integration that will give them maturity. But we must recognize them now, because the stakes are the most clear and the involvement the most dramatic. Later on the need for organized action, political influence and bargaining will blur the passion of revolt and conceal the discovery of the nature of the new conflicts.

Identity and Responsibility

The renewal of social movements, the questioning of the cultural orientations of industrial society, the transformations of the ruling class, all are structural changes that combine to create a new type of society, defined as a technological and social system, directed not by values or forces, but by social relations and political processes.

The old separation between the level of "events," the level of the ruling class and political power, and the level of social and cultural structures (kinship systems and local communities, religious ssytems and agricultural methods) is replaced more and more by the incorporation of the vast majority into mass production and consumption, by an extended participation in political mechanisms, and by the influence of active or passive changes.

Instead of the pair: domination of a metasocial order and cultural identity comes the pair: system and conflict. The actor can no longer be defined by his statuses and roles in a community but by tensions, conflicts, and cultural and social changes he has to live through and by his revolt against domination, which becomes broader and broader, covering itself with a mask of rationality.

If this is the nature of the social actor, is not sociology straying out of its main track when it reduces him to statuses and roles?

Instead of studying economic attitudes and behavior of richer and poorer, of managers and workers (an operation no less superficial when it pretends to be a study of classes), we must get away from the actor and from his subjectivity and analyze social relations, the system of orientation, adaptation, and organization of society, before coming back to the actors and observing the tensions and conflicts of a society that is not only reproducing itself but that changes and is able to create its own field of experience and its own orientations. This break with positivism is the more difficult when it is constantly changing its appearances.

If we are ready to refuse to consider society as the embodiment of values or natural needs, is it not tempting to define it as the incarnation of a ruling ideology? Beyond the differences between their political connotations, however, these two formulations are not very far apart and both lead to the dissolution of sociology, making it impossible to understand how social organization and social change are determined by the action that society exerts on itself and by class conflicts through which action then takes shape.

Alain Touraine

What sociology needs most today is a redefinition of its object. Not of its ideas, but of the object it builds, of the social image of the social actor. A meditation on identity is a preliminary step in a search for new forms of social action.

Protest and Change

The Death of Mishima*
Donald Keene
Columbia University

Specialist in Japanese literature and culture. Professor of Japanese literature, Columbia University, since 1955. Author of The Japanese Discovery of Europe *(1952),* Anthology of Japanese Literature *(1955),* Modern Japanese Literature *(1956),* Living Japan *(1959),* Major Plays of Chikamatsu *(1961),* Modern Japanese Novels and the West *(1961). Contributor to Japanese publications and the* New York Times Magazine.

* Copyright © 1971 by Donald Keene

Yukio Mishima, minutes before his cry of "Long Live the Emperor!" and public suicide on November 25, 1970, in Tokyo. Courtesy of Yomiuri Shimbun.

Yukio Mishima's suicide, under spectacular circumstances, on November 25, 1970, attracted extraordinary attention throughout the world. In Japan, naturally, interest ran highest; every man who had ever sat next to Mishima in elementary school or nodded to him in a corridor at the university felt impelled to describe unforgettable memories. Sometimes (especially before the television cameras) old friends even shed tears at the thought that if only, thirty years earlier, they had shown more comradely interest in the lonely, withdrawn young Mishima, the terrible tragedy might have been averted. There were also the scandalmongers, for the most part anonymous, who with a leer (or an air of "now it can be told") eagerly related unsavory incidents alleged to have occurred in Mishima's past. Others, unmoved by any personal relationship with Mishima, denounced him as a fascist bent on a revival of Japanese militarism with all the terrible consequences of the 1930s. Others still, including students of the extreme left, praised him as a man who carried into action his disgust with society, unlike the usual liberal intellectuals, whose indignation never goes beyond empty pronouncements. Such praise from adversaries was, in turn, balanced by criticism from his closest associates, the members of his private army (the Shield Society, since disbanded), who denounced Mishima for betraying the ideals of the organization in order to indulge in a private gesture. It has been rare to find a literate Japanese without strong opinions on Mishima's death.

Japanese publishers, needless to say, have enormously profited by this absorption with Mishima. Those with works by him on their lists, even ephemeral recollections of the 1964 Olympics or plays in eighteenth-century verse, were quick to reprint them, and sold many copies. Special sections were set up in the Tokyo bookshops for works by Mishima, and there was a flood of books and magazines devoted exclusively to him, containing essays by both friends and enemies. Collectors of first editions of Mishima must now expect to pay upwards of $500 for choice items with original dust jackets, and a specialized study of Mishima rarities even lists the instances when a second edition is scarcer, and therefore more expensive, than the first.

Donald Keene

It would be hard to think of any event since the end of the war that has exerted a stronger impact on the Japanese public, despite the policy adopted by the major newspapers of playing down Mishima's death, for fear the details might adversely affect the impressionable. Yet, if his suicide was intended to stir the Japanese into action, it has clearly been a failure. Mishima was ostensibly attempting by his suicide and his last exhortation to the Self-Defense Force to awaken the soldiers to the ambivalence of their position as members of an army that is not an army, the establishment of a Japanese army being prohibited by the present constitution. Mishima was jeered at by the soldiers throughout his speech, so loudly that the recording made at the time is almost inaudible, and the soldiers seem to have had no second thoughts. Undoubtedly some senior officers nostalgically approved of an act so reminiscent of the spirit of the old army, but the rank and file, whatever their personal opinions of Mishima, have dismissed his act as an anachronism smacking of an era with which they do not sympathize. The magazines aimed at the Self-Defense Force have been at particular pains to print articles discrediting Mishima, in the effort to make his actions seem less a challenge to the soldiers than the working out of a private and perhaps contemptible dilemma. If anything, Mishima's suicide has made revision of the constitution more difficult.

Regardless of Mishima's proclaimed intent, many people have interpreted his suicide as a protest against the existing state of Japanese society. The country is more prosperous than ever before in history, and the economy is the object of wonder and envy everywhere. The elevator boys in Rome, the souvenir shop owners in Cairo, and the diamond merchants in Amsterdam are now likely to speak a little Japanese, though this was inconceivable twenty years ago, so eager are they to have their cut of Japanese affluence. And not even the most radical Japanese intellectuals are untouched by the orgy of acquisition, whether of electronic equipment or of foreign-made clothes. The present, conservative government and the Communist Party both seem to have no higher goals than a Japan totally free from poverty in which every family has a color television set and a car.

But, Mishima wondered, is that really enough? Already Japan is clogged with cars that discharge clouds of carbon monoxide into narrow streets never meant for such traffic. On Sundays in the summer an unbroken stream of cars extends all the way from Tokyo to Karuizawa in the mountains some ninety miles away, and the chief sight awaiting the happy motorist and his family, at the end of a grueling seven hours' drive, are a few tourist attractions built atop the remains of an eighteenth-century volcanic eruption: a garish, completely spurious Buddhist temple, a concrete and glass edifice dispensing cheap lunches and souvenirs, and photographers' stands where visitors are posed against the wasteland of lava.

Mishima's disgust with such postwar "improvements" in the Japanese way of life and with Japanese proclamations of self-satisfaction over the Gross National Product and other evidences of prosperity was deep-seated. Above all, he considered this emphasis on materialism to be un-Japanese. Affluence in Japan was traditionally subdued, or even masked in the guise of honorable poverty, and display was rarely an end in itself. Mishima, believing that the corruption of Japan had originated with superficial imitation of the West, gave many of his writings a seemingly xenophobic streak; the foreign characters in his novels are almost uniformly unattractive. *The Temple of the Dawn*, the third volume of his tetralogy *The Sea of Fertility*, includes such typical scenes of postwar Tokyo as the large garden party at which foreign women, grotesquely made-up and costumed for their age, greedily snatch appetizers from a tray, cackling all the while in their penetrating voices. It does not take a Japanese to feel irritation at the sight of elderly American tourists in gaudy "playclothes" enjoying themselves in Tokyo as if it were another Honolulu. But there was a difference between the reactions of even the most sensitive foreign observers and Mishima: they deplored the bad taste of the visitors, but he acutely feared that his country was being denatured, not so much by the braying visitors themselves as by their

Japanese admirers and imitators, especially the politicians who professed foreign ideals of material well-being, to the exclusion of all the traditional Japanese ideals.

Mishima turned with nostalgia and even envy to accounts of the Japanese fanatics of 1876 who desperately fought against the modernization of the country, challenging the artillery of the government troops with swords and spears. The most extreme among them refused even to walk under telegraph wires, for fear of being polluted by the foreign inventions, or to light their hearth fires with impure foreign matches. These men, fighting against hopeless odds, were massacred, but Mishima considered that their suicidal gesture was truer to Japan than the calculations of the industrialists.

It is easy, comparing these beliefs of Mishima with his private life, to note the contradictions. Despite his antiforeignism, he probably had more and closer foreign friends than any Japanese writer, and his tastes in art, food, and even literature tended to be un-Japanese. Such contradictions have puzzled his Japanese admirers. Should a man who wrote with such profound admiration about the practitioners of the pure Japanese Way have lived in a Spanish-style house furnished entirely with European objects, down to the large marble statue of Apollo in the garden? Should Mishima, despite his rejection of the mindless Japanese adulation of the West, have been so responsive to the changes in men's fashions, going in for sport shirts that opened almost to his navel or for trousers so tight they became uncomfortable after he ate? For Mishima these surface contradictions had no significance. It made no difference whether he ate a steak or raw fish, and if he preferred a European oil painting to a Buddhist scroll, he was sure that this did not touch the core of his beliefs.

Mishima gladly accepted the luxuries his large income permitted, but he understood and even sympathized with the younger generation in Japan or America who rebelled against such affluence. In an interview recorded just a week before his death he recalled how he had offered two years earlier to join hands with the students occupying the buildings at

The Death of Mishima

Tokyo University, provided they shouted with him, "Long live the Emperor!" This condition was, of course, unacceptable to the students. For Mishima this shout was a declaration that beyond the destruction of the existing order, to which he consented, there must be something worth dying for and a readiness to die for it. Mishima expressed in the interview his contempt for the students who in the end meekly surrendered to the police, rather than leaping from the tower of Tokyo University, but he may secretly have been relieved that they had not clung strongly enough to their ideals to die for them; he wanted to believe that worship of the Emperor was the only solution to the meaninglessness of modern Japanese life.

Mishima's worship of the Emperor, his insistence on the infallibility of the Emperor, has suggested to many people that he belonged to the traditional Japanese right wing, but a passage from his last interview indicates the complexity of his attitude. The interviewer, a Marxist writer, said at one point, "It is not that I feel any enmity or animosity towards the Emperor himself or towards the Imperial Family, but I can't help noting that the Emperor system, as a system, is destined always to be used for the purposes of a kind of political power." Mishima answered, "I, on the contrary, do feel animosity towards the Emperor himself. I reject all his acts of humanization since the war."

His Emperor-worship had nothing to do with the present Emperor as a human being. Mishima told me once of attending a reception given by the Emperor. He had been dismayed that the Emperor was not a trifle more impressive, but despite his wry comments, the personality of the Emperor was essentially irrelevant to his beliefs. Over-publicization of the Emperor's quirks as a human being (or conscious attempts to create popular feelings of affection, rather than awe, for the Imperial Family) could only weaken the claims of the Emperor to the absolute loyalty of all Japanese. The *kamikaze* pilots who plunged their planes into American ships with a last cry of "Long live the Emperor!" were not dying because they admired the Emperor's expertise in marine biology, nor because they had been touched by his affectionate concern for his people; they invoked his name as a symbol of Japan itself—not the militaristic government, nor the financial empires, but the living tradition of being a Japanese.

Mishima's definition of being a Japanese involved necessarily a readiness to die for the Emperor. He had been attracted ever since he was a boy by the curious work *Hagakure* (Under the Leaves), a kind of textbook of samurai conduct written during the eighteenth century, which contains the classic statement of *Bushido*, the Way of the Samurai: "By *Bushido* is meant perceiving when to die. If there is a choice, it is always preferable to die quickly. There is nothing else worth mentioning." Mishima wrote a popular introduction to *Hagakure* in 1967, and its samurai ideals were undoubtedly in his mind when he cried out "Long live the Emperor!" just before committing suicide. He died for the Emperor, an Emperor for whom he had no particular respect as a person, and who in turn (as far as we can tell) has little taste for such extreme gestures of loyalty. His avowed object, stated in the manifesto he distributed, was to arouse the Japanese to an awareness of the hypocrisy and corruption in their lives.

Mishima's convictions seem to have led him inexorably to suicide, but he needed some direct cause for the act. In his final address he declared that he was killing himself because of his disgust over the failure of the Self-Defense Force to assert its reasons for existence. He shouted, "What is the point of the Self-Defense Force becoming an army? It is to defend Japan. Yes, to defend Japan. And to defend Japan means defending the traditions of our history and culture, centered around the Emperor." At this point the jeering became so loud Mishima cried out in exasperation, "Listen to me, listen! Shut up! Stop that noise! Listen to what I have to say! A man is appealing to you, staking his life on it." But the soldiers did not listen. Their raucous shouts grew louder, and nobody expressed approval when Mishima declared that unless the soldiers acted to revise the constitution, the Self-Defense Force was doomed to perpetual bondage to America. "Isn't there even one of you willing to stand with me?" he asked, and paused five seconds. "No, not even one! All right, then . . . If you still won't stand up in favor of changing the constitution, I know what that means. My dreams for the Self-Defense Force are over. I'm now going to shout, "Long live the Emperor!"

These were his last words. It is hard to doubt the sincerity of words pronounced a few seconds before a suicide. But did Mishima's speech really express his reasons for killing himself? Did he really entertain any illusions about the soldiers rising and joining him? Surely his long association with the Self-Defense Force should have plainly revealed to him the general indifference of the men to political issues, and even their reluctance to admit they are soldiers. Mishima could not have hoped that at best more than a handful would support him.

Then, why did he do it? Mishima's suicide surely cannot be dismissed as merely a stunt to attract public attention. Undoubtedly the indignation expressed in his manifesto accurately reflected his disillusion with the sterile banality of Japanese prosperity. But there was much more to Mishima than his role as a critic of Japanese society. Shortly before his death he staged at a Tokyo department store a massive exhibit of photographs and memorabilia of himself, dividing his life into four "streams"—literature, the theater, the flesh, and action. His manifesto explained his death in terms of public convictions, but undoubtedly each of these four "streams" also led him to the same decision.

As a writer Mishima had achieved the highest distinction. Not only was he generally recognized as the best Japanese novelist and playwright actively engaged in writing, but translations had spread his fame to many countries. No doubt it was a disappointment when Kawabata was awarded the Nobel Prize in 1968, thus precluding his own chances at the prize for years, until Japan's turn came round again. Kawabata, with characteristic modesty, stated at the time that Mishima, a genius whose likes were seen only once in 300 years, deserved the honor more than himself, and perhaps he was right. Mishima certainly wanted this international cachet of approval. The importance of foreign evalu-

Donald Keene

ation of his works is revealed by the prominence of the requests in his last letters for help in ensuring that the English translation of the tetralogy would be published in full. But he must have felt utterly secure about his reputation in Japan. He told me in August 1970 that he had already written enough for one lifetime, and had put into the four volumes of *The Sea of Fertility* everything he had learned as a writer. "When I finish this book I'll have nothing left to do but to die," he said with a laugh, and I laughed too, unable to take him seriously. Surely, I thought, Mishima would soon surmount the feelings of emptiness every author experiences on completing a major work. But Mishima meant it literally. It must have seemed peculiarly appropriate to die on the day that he delivered to the publishers the concluding installment of his culminating work. It was not that Mishima feared a waning of his creative powers. He knew he could go on writing better than anyone else in Japan, almost without effort. But there was nothing more he had to say. He chose to end his career at his peak.

Mishima's career in the theater and films, the second of his "streams," must also have been carrying him towards his last dramatic gesture. Not only did he write plays (and stories) exalting the young officers of the February 26, 1936, revolt against the authorities, contrasting their selfless idealism with the grimy prudence of the old politicians, but he himself appeared in the film he made in 1965 of his novella *Patriotism*. The film, the supreme expression of a familiar theme in Mishima's writings, the union of eroticism and death, portrays the last hours of a young officer and his bride. The officer was not included in the plans of the 1936 revolt because his friends took pity on a newly married man, knowing how likely he was to die if he participated. But he could not allow it to be supposed that he was less absolutely devoted to the Emperor than they. He returns home and informs his wife of his decision to kill himself. She, knowing what being an officer's wife means, does not attempt to dissuade him. They make love for a last time, passionately, then he commits *seppuku*, the ritual disembowelment, and she stabs herself. Mishima performed the *seppuku* scene in an almost unbearably realistic manner, a proof of how closely he had studied the old accounts. Mishima became increasingly fascinated with swords and *seppuku*; perhaps it was the chance to perform a *seppuku* scene again that tempted him in 1969 to appear in another film. This time he took the part of a celebrated swordsman, unjustly accused of killing a noble, who commits *seppuku* to protest his innocence. Having practiced *seppuku* so carefully before the cameras, he was ready in November 1970 to perform it flawlessly in the presence of the horrified general of the Self-Defense Force.

In the film *Patriotism* the lieutenant dies as a result of his self-inflicted wounds. Usually, however, when a man committed *seppuku* a trusted friend stood by to cut off his head and end the agony. Mishima's second-in-command in the Shield Society, Masakatsu Morita, performed this service for him, but so clumsily it took several slashes of the sword before he succeeded in beheading Mishima. Morita's clumsiness no doubt was caused by extreme agitation at having to kill his own mentor and perhaps also by the knowledge that in a few minutes he too would be dead.

Mishima endured the terrible pains without a murmur. His flesh, the third of the "streams," proved equal to this supreme test. Mishima had long told friends, jestingly, that his fanatic cultivation of physical prowess, for which he had become nearly as celebrated as for his writings, was occasioned by an aesthetic consideration—he was determined to leave behind an attractive corpse. No doubt the fear of physical decay was extremely strong. Mishima was eager to die while still capable, though forty-five, of undergoing the punishing training of the most rigorous army life, and even of outdoing the efforts of men half his age. He knew that with each passing year this would become less and less possible, and the thought must have weighed on him. But it was not only fear of growing old and losing his strength that made him choose this most violent form of suicide. From his childhood Mishima had been fascinated by pictures of samurai committing *seppuku*, European knights dying on the battleground, and, above all, St. Sebastian, his powerful body bound to a tree, dying of arrow wounds. Mishima had himself photographed in the pose of St. Sebastian, and there is even a whole album of photographs showing him writhing in weird postures. His narcissism was pronounced, and he never tired of seeing pictures of his powerful body. A masochistic streak, an urge to destroy the flesh, was also apparent in his writings as well as in these photographs. He had made of his body a work of art, only to destroy it with a thrust of the knife. Beauty and death, the two elements in the "stream" of his flesh, both demanded his suicide.

Finally, the "stream" of action was relentlessly urging Mishima to his death. When he founded the Shield Society in 1967 it was with the announced objective of fostering the military virtues and defending the Emperor, but Mishima may have had a more specific objective. Like most other people, he expected that in June 1970 there would be a massive uprising in protest against the renewal of the Security Treaty with the United States. Students and labor groups, it was widely reported, had been organized for action that might even lead to an overthrowal of the government, and had obtained military equipment for this purpose. Moreover, as the riots in 1960 had shown, the public was likely to be indulgent towards demonstrators, no matter how violent, and the only resistance to revolution was likely to come from the unpopular police. This prospect dismayed Japanese intellectuals who were not of the extreme left, but Mishima rejoiced. He anticipated that he and his little army would die defending the Emperor, precisely the death he desired.

But nothing happened. On the day the treaty renewal went into effect, I drove with Mishima past the Diet building, ringed by grim-faced police. He said with a shake of the head that it was hopeless, he was disappointed. I thought this was his usual joke, pretending to enjoy disaster and to be unhappy only in times of humdrum prosperity, but he was far more serious than I realized. As a boy he had hoped to be killed in the war, as he related in his first novel, *Confessions of a Mask*, and he felt cheated when the war ended in 1945 with himself unharmed. He wrote me earlier in 1970 that he had hoped for "a little Goetterdaemmerung." But his carefully laid plans for achieving death defending his Emperor, a samurai's death, were frustrated.

He now had to face the problem of what to do with the Shield Society. If the eighty or ninety members were not destined to die in a gallant last stand, what point would there be in maintaining them? Mishima came to the conclusion that he would have to work out plans for his death without regard for the Shield Society. Already the four "streams" were converging on that point. But Mishima could not kill himself before he completed his tetralogy, the grand climax to his work as an artist, so his "coup" would have to take place after the delivery of the final episodes of the novel to the magazine which was serializing it. Mishima customarily turned over each installment on the twenty-fifth of the month, and he estimated that the book would be completed by November. The date November 25, therefore, suggested itself for the "coup." It would have suited Mishima's purposes if some dramatic incident of a political nature had occurred just prior to this date, providing him with a plausible reason for his *seppuku,* but there was nothing worth mentioning. The most he could say in his final oration was that he had been waiting impatiently, for over a year, for the Self-Defense Force to show its anger over what had happened on October 21, 1969. On that day a left-wing demonstration sparked by Prime Minister Sato's departure for Washington had been squelched, not by the Self-Defense Force, the guardians of Japan, but by the police, who had usurped the functions of an army. Mishima's pretext was flimsy, and the choice of his audience, the amorphous conglomeration of units making up an army headquarters, rather than the disciplined body of a regiment, was dictated by the fortuitous absence from Tokyo of a regimental commander known to Mishima. He was determined to die on November 25, regardless of how inappropriate the circumstances might be.

He could have committed suicide in some other way, of course, but as he told friends in September, he was determined not to die a "stupid" death; he wanted his suicide to have an effect on the Japanese people. If he had taken an overdose of sleeping pills in his study, his action would surely have been interpreted as a private tragedy, like Hemingway shooting himself, and whatever note or manifesto he left behind could not have carried much conviction. Besides, Mishima's fascination with *seppuku,* the traditional death of the samurai, would not permit any less dramatic death. Even once he had decided on *seppuku* he could have done it quietly in his garden, or in the park facing the Imperial Palace, but without the dramatic setting—the commanding general held captive at sword's point by Mishima's men, the assembling of the troops, the oration on the balcony—*seppuku* might have seemed a foolish or even demented gesture.

In order to create the circumstances necessary for a *seppuku* of maximum effect, Mishima took a few of his most trusted followers into his confidence. The survivors of the "coup" would be given the opportunity at their trial to appeal to the Japanese public by restating the ideals for which Mishima died, and a certain number of men were needed in any case to hold the general captive while Mishima addressed the troops. Morita, whom he chose to cut off his head once he had disemboweled himself, would also be permitted to commit *seppuku,* in his capacity as leader of the student members of the Shield Society. It may be that Mishima's conception of the relations of captain and lieutenant, inspired by his readings in *Hagakure,* made it seem appropriate that Morita also die.

As the time approached for the "coup," Mishima's writings became increasingly political. He published in September an article praising the "philosophy of action" of the Chinese thinker Wang Yang-ming (1472-1528), whose disciples in Japan, especially during the nineteenth century, had not been content with quibbling over logical or ethical principles in the manner of most Confucian scholars, but had fought for their beliefs and in the end killed themselves. The tone of his utterances, even though he was engaged in writing his literary masterpiece, became distinctly anti-literary: in September he stated in an interview that he would not accept the Nobel Prize if it were offered to him, giving nationalistic—and rather unconvincing—reasons. His last essay, delivered to a magazine less than a week before his death, was entitled "My Views on Comrades." Its shortness—it is less than three pages long—and its obscure expression suggest the great perturbation in his spirits and perhaps also his inability to concentrate on his subject. Above all, the almost inhumanly political emphasis seems quite uncharacteristic of Mishima. It begins: "I have always thought, with respect to comradely associations, that even if it should happen that a comrade of mine died before my eyes, I would not cling to his corpse weeping, but I would be able, even in a court of law, to affirm that he was a stranger I had never seen before. This is possible only if one maintains the tension of spirit implied by the phrase 'nonemotional solidarity'."

The essay leaves many questions in the reader's mind, but references to the incidents of October 21, 1969, suggest he was already considering his last speech. The left-wing students on that day, he said, revealed their lack of "nonemotional solidarity." They were also unaware that the authorities, who easily put down their demonstrations, would have been helpless if the students had been willing to die. Mishima says he intuitively realized that day that unless men were willing to die the constitution could not be changed for another ten years. He commented, "For persons contemplating revolution it is no doubt natural that they make effective use of death in their strategy and actions as a means of accomplishing their objectives. If there is a guarantee that, at the optimum moment created by one's own actions, death can be employed as a dramatic climax, effectively, this cannot be said to be dying a dog's death."

Mishima's last letter to me, written shortly before his death, stated that he had long wished to die as a samurai, rather than as a man of letters. Undoubtedly this was true, and the tone of his last utterances underlines his resolve to die in the character of a samurai. The long years he devoted to mastering the martial arts, the months when he stopped all literary work to train with the Self-Defense Force, and even the popular books with titles like *For the Young Samurai* bespeak his obsessive interest in the ideals of the warrior.

But Mishima was not a samurai. He was a writer. Of course he was aware of this contradiction, so he felt obliged in the end to prove he was a samurai, despite all evidence to the contrary. A passage in *Five Signs of a God's Decay*, the last part of his tetralogy, may be revelatory of how he conceived his suicide. One character, perhaps voicing Mishima's own views, remarks, "In general, I don't like the weakening, the feebleness of the man who kills himself. But there is one kind of suicide I can allow. It is the suicide intended to legitimize yourself." When asked what this means, he tells the story of the mouse who was convinced he was a cat. One day a cat came along and announced he was going to eat the mouse. The mouse protested, saying cats should not eat other cats, but the cat only laughed. Finally, to prove his point, the mouse jumped into a pail of detergent and drowned himself. The cat tentatively put his paw into the liquid, licked a bit, then decided that a mouse with such an unpleasant taste would not be fit to eat, and went away. The mouse, analyzing what made him a mouse, had decided it depended on two factors, his physical appearance and his natural destiny to be eaten by a cat. He could do nothing to change the former, but by making himself inedible he had proved, at least to his own satisfaction, that he was not a mouse.

It is not clear how closely this fable can be related to Mishima's death, but surely self-legitimization was involved. Years before, disgusted with the unhealthy, harassed look of the typical Japanese intellectual, he had taken up weightlifting, to such good effect that he transformed his frail body into the muscular torso of a Greek statue. In his writings, too, he had rejected with contempt the self-pitying tone characteristic of fiction written by so many novelists entranced by their own suffering, and in his friendships he tended to avoid writers or scholars, preferring the company of unaffected, uncomplicated people. But he himself was an extremely complex man, an intellectual among intellectuals, a writer of exceptional sensitivity and delicacy. Nothing could have been farther from Mishima than the simple, though intense loyalty of the lieutenant in *Patriotism*, but by his suicide he "legitimized" the mask of Emperor-worship he had borrowed from his own creation.

Many contradictions surround the death of Mishima—the *coup d'état*, complete with manifesto, that was not a *coup d'état*, the exhortation to the troops delivered with all seriousness but with the foreknowledge that it would have no effect, the bewilderment of Mishima's own followers over an act that seemed to violate the principles of their organization—but the most basic surely was Mishima's consuming concern during his last days, perhaps during his last hours, over the English translation of *The Sea of Fertility*. Somehow, despite the specific assurances in the contract with his American publishers, he got it into his head that they might decide after his death not to publish the final volumes of the tetralogy. His last three letters—two to friends in New York and one to his wife—asked that each do what he could to assure publication in English of the entire work. But why should a samurai, indignant over the degeneracy of his country, and desirous of stirring his compatriots by the most dramatic and unanswerable gesture of loyalty to Japanese ideals, have worried so much over the English translation of his books? Was this no more than a momentary weakness, a last flicker of interest in his worldly reputation on the part of a man about to give up his life? Somehow I do not think so. The concern seems genuine, of a piece with the consuming curiosity he showed to the last about foreign translations and reviews of his books. He was an artist, a great writer, who found himself driven to suicide as a samurai, but who could not renounce completely his true profession.

Mishima's last article, as I have indicated, is almost incoherent. The manifesto, with its phrases lifted bodily from the manifesto of the army officers of 1936, the two valedictory verses filled with the hackneyed imagery of classical Japanese poetry, and the bluff military style of his oration to the troops do not recall the Mishima I knew. They were part of the terrible mask he had assumed for his last role, but the result was an anomaly. Mishima was trying to legitimize himself as a

Donald Keene

samurai in an age when there is no master for a samurai to serve. He played many parts in his lifetime, all of them with a strange combination of dramatic flair and sincerity, but the mask worn on November 25 was not one I recognized. The Mishima I knew was generous, wonderfully intelligent and witty, easily wounded but quick to forgive, a superb friend. The Mishima the world will continue to know, even after the memories of how he died are blurred, was the writer of a prodigious variety of literary works that do credit to our century.

The Double Revolution
Saul D. Alinsky

Sociologist. For many years an advocate of community action. Executive Director of the Industrial Areas Foundation from 1939 until his death in 1972. Author of Reveille for Radicals *(1946),* John L. Lewis, A Biography *(1949),* The Professional Radical *(1970),* Rules for Radicals *(1971).*

A double revolution is ripping the world, with each acting and reacting on the other. One, global-wide among the have-not nations and in the have-not sectors of the have nations, is the finale of all the revolutions of the past, of the world of the past, the long struggle for physical life, food, shelter, safety, or security.

Materialistic objectives have been and are basic to every revolution of the have-nots, such as the dialectical materialism of Marxism. Materialistic ends are concrete, specific, readily communicated and understood. To the have-nots, goals, values, and truths are fixed, definite, and "self-evident." There is nothing relative about starvation, sickness, or cremation by napalm. They believe as did we in the Great Depression of the thirties that full employment and economic security would carry with them everything else to make up the good life.

So we got the "good life." It came with an unprecedented technological revolution, ushering in a new world of automation, the computer, cybernetics, mass media, nuclear power, the jet, and other undreamt of miracles of production. In the United States nearly eighty percent of our people, "the masses," are middle class; here, the poor are the minority. Here, most of our people are dieting while the have-nots are dying. This is the "affluent society" which beckons on to an economic paradise beyond suburbia, split-level ranch houses, color television, two-car families, a burgeoning corporate economy, and, seemingly, the good life. Now that we have it why are we so unhappy? Why has all this triggered the second revolution: the revolution of the middle class, reaching out, striking out, spinning in a whirl of desperation as though it were in a death agony?

Why, with this economic floor, was America so recently tearing itself apart with violence, campus rebellions, racial upheavals, a war which revived Talleyrand's classic comment, "This is worse than a sin; it is stupid."

Inflation and recession all combine to the point where we are confused, disunited, depersonalized—not only from each other but from the society around us—alienated and fractured with "gaps" of every kind, everywhere, from generation to credibility to communication and everything else.

The rhetoric of hate, death, destruction, and smear of dissent from the so-called far New Left was indistinguishable in its madness from that of President Nixon's former political mate, his vice president, who pronounced Vietnam as America's most moral act in foreign policy; or the fetid retching telephone calls of a former attorney general's marital mate.

We are lost and frightened. We don't know where we are and we are scared to death of where we may be going or not going. Scared New World.

We are in times as different from the past as the computer is from the abacus. We see a frantic searching for reasons for our fears, our unhappiness, for some meaning to our lives. We see a turning away inward into a social schizophrenia where, in a flight from frustration, fear, and bewilderment, people seek not to be "involved"—a turning away from life itself for life *is* being involved. Not being "involved" means dropping out as completely as the hippies, or those on drugs. The only difference is that their dropout is conspicuously visible. This is the real gap, the one between the people and the new world they live in. This and other gaps create huge political vacuums which bode ominously for an open-society future. Political vacuums produce and bring to power totalitarian demagogs.

We have seen the shadows of possible things to come in the campaign speeches which read like this:

> The streets of our country are in turmoil. The universities are filled with students rebelling and rioting. Communists are seeking to destroy our country. Russia is threatening us with her might and the Republic is in danger. Yes, danger from within and from without. We need law and order. Yes, without law and order our nation cannot survive. Elect us and we shall restore law and order.
> —Adolf Hitler
> Hamburg, Germany, 1932

All this and more spells out the forces now loose in our middle classes.

We must begin to try to make sense of the whys and wherefores of our confusion. Until we have an idea of what is happening, know where we are, we cannot choose a direction, a way, or any meaningful action for the future. We must cut to the core of what this revolution of the middle class is about. It is a new kind of revolution, as new as the whole new technological world of today. The clues are to be found in the plaintive cry of "Where have the old values gone?" All of the values of the past were fixed, definite, and final; good was good and evil was evil. We see it in the revolt of the have-nots, with their fixed, definite values of economic security, good housing, success, status, and the faith that these will bring them happiness.

What happened to our values? It began in the field of physics, and the names of the revolutionaries who unwittingly blasted the world that we had known are those of Max Planck, Albert Einstein, Max Born, Pascual Jordan, Niels Bohr, and Werner Heisenberg. From Planck's creation of the quantum theory to Einstein's relativity, to Niels Bohr's complementarity and on to Heisenberg's principle of uncertainty, the field of physics produced two bombs: the atom bomb to destroy physical life and an even more devastating political bomb of relativity, which blew apart the world of the values of the past. It ended a world of fixed absolute values, or formulas, of everything operating from precise cause to precise effect; it ended a world of definitive certainty, and in its stead ushered in a whole new world of values wherein everything is relative, changing, and operating in terms of probabilities rather than cause and effect—and in which the only certainty is uncertainty. And so the young generation is denouncing the present society for its "bankrupt, degenerate, decadent, materialistic, middle-class, bourgeois values" and demanding "new values" with much revolutionary rhetoric. The older generation is deploring the fact that "the values that they grew up with are gone and the world is going to hell."

It is when we begin to understand that we are now living in a world in which everything is changing and relative that we can begin to perceive dimly and try to move with these relative and changing values. It means that programs, plans, and systems of ethics cannot be rigid and structured but fluid and flexible, going with and shaping themselves with the action. Space does not here permit more than these brief comments on the fountainhead of the middle-class revolution. Suffice to say that if the Founding Fathers were writing the Declaration of Independence today it would read, "We hold these truths to be self-evident and relative," for it is today undeniably self-evident that all truths are relative.

We are crossing the chasm from a world in which security was found in definite permanent truths, values, and goals, to a world of constant change and relativity. The fact that the past view of fixed, finite values and formulas and the idea of possible security was an illusion is irrelevant; we believed them to be reality. It would be expecting the impossible that one or two generations could bridge the chasm from a life of illusion to one of reality, accept and adjust to a world where uncertainty was the one certainty, and all values, ethics, and goals were relative and constantly changing. It is in many ways man's first great confrontation with reality; more revolutionary than man's getting up from his four legs and walking on two would be his walking without the crutches of illusions.

In America the arena for action and change is in our middle class. Marx never foresaw a scene where three-fourths of a population—from self-identification and economics— is middle class; where the proletariat constitutes a distinct minority; where even a coalition of organized units of blacks, Mexican Americans, Puerto Ricans, Appalachian whites, and other low-income sectors would be insuf-

THE NEW YORK TIMES, MONDAY, JUNE 26, 1972

Friends Say Good-by to Alinsky: Kaddish, Drinks, a Few Laughs

By JOHN KIFNER
Special to The New York Times

CHICAGO—They said the Kaddish—the Hebrew prayer for the dead—for Saul David Alinsky, and then his friends went to the back room of a bar named the Boul' Mich and told stories and laughed, which was what he would have wanted.

There was Roman Catholic priests and black activists among those remembering the "professional radical" and organizer of the poor and powerless; young whites whom he was training to work in middle-class neighborhoods; newspapermen, and old-timers from the days of organizing the packinghouse workers in the Back of the Yards.

The memorial service was held at the Kehilath Anshe Maarav Synagogue near the University of Chicago, last Monday. Mr. Alinsky died of a heart attack in Carmel, Calif., on June 12.

In his eulogy, Msgr. John J. Egan, who helped to bring the support of the powerful Chicago Catholic archdiocese behind his projects, summarized Mr. Alinsky's motivation.

"Jack," he recalled Mr. Alinsky saying, "I just can't stand to see people pushed around."

Pushed Back

He began pushing back in the late nineteen-thirties with the formation of the Back of the Yards Neighborhood Council, the prototype of what he called People's Organizations, and in 1940 started the Industrial Areas Foundation.

In the late 1950's Mr. Alinsky built The Woodlawn Organization in the black neighborhood around the University of Chicago, and he and his organizers moved into other cities—Rochester, Buffalo, Kansas City and Detroit—to build power blocs of the poor.

In recent years, he turned much of his attention to organizing projects among the white middle class which, he felt, had become powerless and alienated, and tried to form alliances between white and black groups.

Mr. Alinsky's tactics were rough—"If the end doesn't justify the means," he once growled, "what the hell does?"—and often involved ridicule that would outrage city officials.

'I'm Having a Ball'

He expressed his creed in his book "Reveille for Radicals," which was published in 1946. Much of it had been written in the relative peace of a Kansas City jail. A blend of Jeffersonian faith in democracy and shrewd, sometimes cynical, maneuvers, it became a manual for community organizers.

"You know what it is about me that really burns people?" he asked after one bruising and successful struggle. "It's because I'm having a ball."

After Mr. Alinsky's death, one of the independent aldermen in the Chicago City Council offered a motion to name a park after him. The motion was quickly defeated by the Administration's usual 35-to-8 majority. His friends thought that he would have had a good laugh over that.

ficient to affect basic changes. The power lies in the vast middle-class mass. It is here where the die will be cast. Here, the so-called silent majority will determine whether man will continue living in what Thoreau described as "quiet desperation" and dreams of a free and open society will continue to erode into a closed dictatorship. We may well become the first totalitarian state with a national anthem about "the land of the free and the home of the brave," but then I must sadly concede that even the concept of freedom is relative.

America's middle class is scared silent. Reacting ostrich-like to the tragedy, terror, and senselessness about us, but unostrichlike and like humans always looking for hope or rationalizations, it cannot rest by burying its head but burrows it backward, emerging to look up at its rear end and believing, because it wants to, that it is the beatific vision. How else can we rationalize the grotesque absurdities of the times?

We are the age of pollution, progressively burying ourselves in our own wastes. We announce that our water is contaminated by our own excrement, insecticides, and detergents—and then do nothing. Even a half-sane people would long since have done the simple and obvious—ban all detergents, develop new nonpolluting insecticides, and immediately build waste-disposal units. Apparently we would rather be corpses in clean shirts. We prefer a strangling ring around the neck to a "ring around the collar." Our persistent use of our present insecticides may well insure that the insects "shall inherit the world." The irony of clean cadavers or man's ignoble death through insecticides or sinking in our own toilet waste is equalled or topped by the absurdities of everything else.

Of all the pollution, water or air, none compares to the political pollution of the Pentagon. Here the utter madness and stupidity make the story of Dr. Strangelove a credible, rational understatement. From a war simultaneously suicidal and murderous in Vietnam to a policy of getting out by getting in deeper and wider; to the Pentagon reports that strained even a moron intelligence with their repeated claims that "within the next six months" the war would be "won"; to destroying more bridges in North Vietnam than there are in the world; to the astronomical counting and reporting of enemy dead from helicopters, "Okay, Joe, we've been here for fifteen minutes, let's go back and call it 150 dead"; to brutalizing our younger generation with My Lais but ignoring our own principles of the Nuremberg trials; to putting our soldiers in conditions so conducive to drugs that we stood forth as freedom's liberating force of pot and mercenaries. A Pentagon whose economic waste and corruption have been bankrupting our nation morally as well as economically. A Lockheed Aircraft that put one-fourth of its production in the small Georgia country town of the then-influential, former military-appropriations senator, and then successfully maneuvered our government to dump federal millions into it to save it from its financial fiascos. Similarly, the situation that prevailed in the congressional district of the late congressman who chaired the House Military Affairs Committee. His district got phenomenal payoffs of every kind of installation from corporations vying for Pentagon gold. This corrupt tragedy was amusing to our unbelievable former vice president.

. . . Vice President Agnew praised Mr. Rivers for his "willingness to go to bat for the so-called and often discredited military industrial complex" as 1,150 generals, Congressmen and defense contractors applauded in the ballroom of the Washington Hilton Hotel . . . .
. . . Mr. Agnew said he wanted "to lay to rest the ugly, vicious, dastardly rumor" that Mr. Rivers, whose Charleston, S.C., district is chock full of military installations, "is trying to move the Pentagon piecemeal to South Carolina."
"Even when it appeared Charleston might sink into the sea from the burden," said the Vice President, Mr. Rivers' response was, "I regret that I have but one Congressional District to my country to— I mean to give to my country."
—*The New York Times*,
August 14, 1970.

This is the Pentagon that has manufactured nearly 16,000 tons of nerve gas, to over kill the overkill. No one has raised the questions as to who got the contracts or what it

The Double Revolution

cost or where the payoffs went. Now the big question is how to dispose of it as it deteriorates and threatens to get loose among us. Forgetting the cost of disposition, we have a scene where the Pentagon announces that the ocean-sinking of the nerve gas is safe *but from now on they will find a safe way!* The obvious American way of assuming personal responsibility for one's action is utterly ignored. Since the Pentagon made the nerve gas, it should keep it—and have it all stored in the basements of the Pentagon; or, since the President, as commander in chief of our armed forces, believed that the sinking in the ocean of the sixty-seven tons of nerve gas was so safe, why didn't he publicly attest to his belief by having it dumped into the waters off San Clemente, California? Either action would at least have given some hope for the nation's future.

The record goes on without any deviations toward sanity, as witness the army's selection of the final day's hearing of the President's Commission investigating the National Guard killings at Kent State to announce that M16 rifles would now be issued to the National Guard! This would insure far more than four killings on the next round. The President's Commission report was doomed not to be read until after the football bowl games on New Year's Day by a President who watches football on TV the afternoon of the vast Moratorium Day march in Washington. There are our generals and their "scientific" gremlins who, after assurance of absolutely no radioactive menace from the atomic tests in Nevada, were forced more than a dozen years later to seal off 250 square miles as "contaminated with poisonous and radioactive plutonium 239." (The *New York Times*, August 21, 1970). This from the tests of 1958! Will the "safe" disposition in 1970 of the nerve gas still be as "safe" a dozen or less years from now? One can only wonder how they will seal off some 250 nautical miles in the Atlantic Ocean. We can only assume that these same "scientist" gremlins will be assigned to the disposition of the thousands of tons of additional stockpiled nerve gas.

Compound this with what was recently a daily record of "now we are in Cambodia, now we are out; we are not in it

Saul D. Alinsky

just over it with our bombers; we will not get involved there as in Vietnam but we can't get out of Vietnam without safeguarding Cambodia; we are escalating the war; we are bombing Hanoi and Haiphong; we are de-escalating; we're doing this but really the other;" with no other clue to all this madness except the half-helpful comment from the White House, "Don't listen to what we say, just watch what we do," half-helpful only because either statements or the actions are sufficient to make us freeze into bewilderment and stunned disbelief. It is in such times that we are haunted by the old Greek maxim, "Those whom the gods would destroy, they first make ludicrous."[1]

If our people are to have "the ability to act,"[2] to change our course from that of headlong self-destruction, then they must organize. Organization does not come into being spontaneously or as a consequence of an immaculate conception; it must be fertilized by a catalyst, activist, outside agitator, radical, or whatever other labels are used. The hope for the future lies in its generation of radicals. Here, too, we are confronted with a bedlam of frenetic impotency.

Part of the cause for the sterility of our radical movement and the former far or New Left, which was so far out as to be better understood as part of our space program rather than the political scene, has been in the breach of the radical experience continuum. In the past the radicals of each generation passed the torch of experience to the next generation. The late Senator Joseph McCarthy's political inquisition of the early 1950s destroyed most of the radicals of that generation, and of the surviving remnants most continued living in the past, mouthing the pretechnological society's now irrelevant revolutionary formulas. Thus, the recent radical generation was aborted into a world without radical parents or guides. Is it any wonder that they cooped themselves into a chronological cage, rejecting anything over thirty years of age, from persons to history.

Power for change rests in the middle class, and there is where young radicals should organize. Inexperienced and unversed in the politics of change, they reject one of their most invaluable assets, a middle-class background with familiarity with middle-class values and way of life. Instead of denouncing the latter as "materialistic," "decadent," "bankrupt," "degenerate," "parasitical," "bourgeois," and "immoral," any political sophisticate knows that communication can only be done within the experience of the subject, and familiarity with the middle-class experience becomes a treasured essential asset to any radical.

When I refer to radicals I do not include those whose actions are rhetorically revolutionary, but result in building strength for the reactionary far right; that is, groups such as the Weathermen, whose psychopathy results in reactions which not only reinforce the right but also open the gates to suppression of freedom in the name of law and order. The bombing, the sniping at police, the Marin County courtroom shootout, the glorification of the assassination of the late Senator Robert Kennedy as revolutionary acts were typical of this horribly sick cult. Any real revolutionary party would have executed them as dangerous counterrevolutionaries. Such actions torpedo that reformation of large masses so essential as the prelude to revolutionary change. Part of what has happened is just straight criminal action in search of a political rationale, such as the absurdity that "all black prisoners in every penitentiary are 'political prisoners.' "

Another alleged activist species I do not include as radical is that substantial segment suffering from a malaise masochism, who, for reasons of their own, desperately seek martyrdom and punishment; and those consumed with a suicidal obsession (always snarling, "Let's get guns and die like men"). Both score zero in any test of political sophistication and effectiveness. The following description of a Yale University student meeting during the height of the New Haven Black Panther episode in the spring of 1970 is a perfect and almost unbelievable example. (The italics are mine.)

.... rage and pandemonium ... prevailed in the first student meeting held on the evening of April 15, the day after Hilliard and Doublas were jailed for contempt. Among the milder courses of action suggested: occupy Woodbridge Hall, *kidnap Kingman Brewster, shut off the New Haven water supply.* A more direct tactic was offered for the imminent May Day rally by Tom Dustou, a Yale dropout and a chief coordinator of the Panther Defense Committee. *'Give me some money and I'll buy guns!'* he screamed, *'I'll stand at the Green and distribute them!'*

An exotic proposal was brought to the floor by an overwrought law student, who suggested mass suicide: Let each person there be allotted a number, he said, and then each day for the following month the person whose number was drawn would give up his life in support of the Panthers.

'Why die?' a student piped up in the stunned silence that followed. *'To die like a Panther, to die like a man,'* the future barrister cried. The meeting dissolved in an apotheosis of radical guilt and threats of burning down the university, with a vote for a three-day moratorium of classes and a demand that the Yale Corporation donate $500,000 to the Panthers' legal defense fund.[3]

Would you believe it!

Yet the beauty of that generation is that the big majority of the activists are now very rapidly becoming politically educated and starting to get into organization—to build power, recognizing that there is no shortcut. They have accepted what is the hardest lesson for youth: that you can't have instant change and that you must begin from where you are.

They are recognizing that communication is basic to organization. It is paradoxical in these days of mass media that we are experiencing an unprecedented breakdown in communication. Everywhere, we are confronted with gaps—generation, credibility, communication. It is not only the collapse of communication between the younger and older generations in the middle class or between the lower middle class and the middle middle class, but also in the areas of race and politics. It seems that all channels of communication are constipated.

In the field of race relations we have no communication. In our present kind of situation, if a white person makes a statement so outrageous that other whites' response is, "you should go to Bellevue for a spinal test," the same statement made by a black draws from the same whites the response, "Well, that's an interesting approach."

Or, the kind of episode such as the one in which the head

of a leading real estate agency in the city of Chicago, who has devoted years to the battle for racially integrated residential housing, was attacked over the air by a well-known, nationally publicized black spokesman in Chicago with the irresponsible and false charge that his real estate agency had engaged in panic peddling. This equal-rights real estate maverick was stunned but did not respond to the attack. I asked him whether he would have remained silent if his accuser had been white? His reply was an angry, "Answer him? I'd have sued the hell out of him—for libel, slander, and everything else, and exposed him for a lying bastard!" I pushed with, "Then why don't you? Is it because he's black?" There was a silence and then his words came slow and tired, "Yes, and of course I suppose my being white automatically makes me a liar to anyone who is black."

So long as this condition prevails, so long will there not be any meaningful communication or constructive positive changes for a world of equality. Many whites have become terrified of even raising a question with blacks for fear of having it branded as a race issue. Unless whites overcome their own hangups so that they can both listen and speak to blacks in the same way that they would be listening and speaking to whites, and vice versa, we are faced with an imminent period of a few years wherein a combination of black charlatans and white neurotics will sow a scene of disillusionment and bitterness which will provide a comforting rationale for all racial bigots, both black and white.

The breakdown in communication is in the political as well as in the color spectrum.

Not too long ago, at an eastern university, a representative of the far New Left concluded a negativistic appraisal of our society as competitive, aggressive, and hostile as compared to the goals of a socialistic society rooted in collective cooperation and love, concluding with a series of statistics including the "statistical" fact that the United States had the largest police force of any nation in the world, far more than that of the U.S.S.R. or China. My inquiry as to his source of information concerning the number of police in the U.S.S.R. or China resulted in an angry glower and then a venomous,

Saul D. Alinsky

"You are red baiting!" The smear, always the answer of the far right to a question, now comes from the far, far left.

I am reminded of Socrates' decision when he took the hemlock that life afterward would either be nonexistent or, as he put it, "a dreamless sleep," or he would find himself in another world able to talk with people like Homer or Orpheus or, as he put it, "the kind of world where one is not killed for asking questions."

The organizer should never confuse realism with cynicism. Oscar Wilde's definition of a cynic, "One who knows the price of everything and the value of nothing" is the antithesis of the organizer who is committed to human values.

As realists, we must see and work with the world as it is and not as we would like it to be. This means that we begin from where we are and begin working for those changes that can come from *revolution* not *revelation.*

This includes shedding romantic views glorifying the poor and the blacks, browns, yellows, or whites. Poverty is ugly, evil; it is survival by sufferance, but the fact that the have-nots exist in rot, degradation, discrimination, deprivation, and despair does NOT endow them with any special qualities of charity, justice, wisdom, mercy, or nobility. They are people with all the faults of mankind.

It means understanding that in the world as it is, where the right thing is invariably done for the wrong reasons, morality is the process by which the right reasons are dredged up to justify the action.

Morality is to a major degree a rationalization of an individual's position on the power pattern at a particular time. If he is a have-not and out to get what he doesn't have, he appeals to a law higher than man-made law and argues that the establishment made the laws to protect and preserve the status quo. If he is a have and out to keep what he has, he believes in law, order, and responsibility (defined, like all words, by where you stand). It is what I have often called the MPI Formula, wherein *M*oney *P*rincipal *I*nvestments=*M*oral *P*rinciple *I*nterests.

In the world as it is, organization is based upon self-interest, specific, immediate, and realizable. The ultimate

The Double Revolution

goal is the learning of the interdependence of man upon man, and that our self-interest ultimately lies in the area of the general social interest. This has been generally described as "enlightened self-interest." In the past my experience has been that this educational process only functions when the self-interests of various groups are mutually dependent upon each other at particular times, or when the achievement of a particular self-interest necessitates concern for others. One example would be a community organized to eliminate a contagious disease within its area, which soon discovers it must work for the same purpose in its neighboring community, since germs are illiterate and don't know that they're not supposed to cross over boundaries into communities organized to keep them out.

Experience and political realism question the validity of the potentials of self-interest developing and expanding into "enlightened self-interest." This assumes that through experience one inevitably realizes that one's personal welfare is inextricably tied to the welfare of one's fellow man. Ordinarily, one would be compelled to concede that from past experience the acceptance of the idea that self-interest would lead to the "enlightened" stage just hasn't worked.

The unprecedented pace of the cascading changes of the times, however, and the immediacy of information have created a condition which is collapsing the time factor—where the future is increasingly intruding upon the present so that the time factor as we know it will soon be part of the old world. The feedback of the consequences of your acts is so immediate that the future is now. We have stepped up the time future as well as pushed up the understanding of the power of the past so that they are all simultaneously functioning in the fleeting instant of what we call the present. It is a new world.

Today we hear the perennial question: Is there time? Can we do anything as the world seems to be hopelessly heading for extinction? Why keep getting into the arena? Why keep fighting? Part of the answer is in the words of W. B. Yeats in *On Tragic Joy:*

Saul D. Alinsky

> *We begin to live when we*
> *conceive of life as*
> *tragedy and understand it to*
> *be truly and persistently*
> *tragic. The fortunate ones*
> *are those who recognize*
> *with a strange joy the terror*
> *of existence. With a gaiety*
> *transfiguring all that dread,*
> *they learn to love life because*
> *it is life.*

We must believe in man's struggle for an ever better world; that man is moving toward a world of more beauty, love, laughter, and creation. That is the vision of man.

Logic and faith go together as the opposite sides of the same shield. We know by our intelligence the greatness and desirability of a free and open society over all other alternatives. Logic tells us, 'We'll believe it when we see it.' But there is also the converse, faith. Faith, or belief in the people, tells us, *'We'll see it when we believe it.'* [4]

NOTES

[1] SAUL D. ALINSKY, *Rules for Radicals* (New York: Random House, 1971).

[2] Definition of the word "power," Webster's Unabridged.

[3] FRANCINE DU PLESSIX GRAY, "The Panthers at Yale," *New York Times*, June 4, 1970.

[4] SAUL D. ALINSKY, *Reveille for Radicals* (New York: Random House Vintage Press, 1969), page 235.

Radical Alternatives to Schools
Ivan Illich
Centro Intercultural de Documentación, Cuernavaca

Educator. Founder and director of the Intercultural Center of Documentation in Cuernavaca, Mexico, an organization of scholars engaged in the study, analysis, and publication of sociocultural information about Latin America. Former vice president of the University of Puerto Rico. Author of many critical studies of contemporary education, including De-Schooling Society *(1971).*

The success of a symposium usually depends much more on surprising circumstances than on the greatest effort of its organizers, and had I been able to listen earlier to Mr. Conor Cruise O'Brien, I should have certainly spoken differently because I, too, will discuss the symbolic—the ritual—the theatrical scenery of man as the actor. I want to talk about the breakdown of a particular stage, because we are about to transform it into an asylum—namely, the school. But, since this, too, is a stage, I had better not react as I would like to—spontaneously, to a previous speaker—but stick to my prepared text.

One of the most interesting things in our epoch is an extreme disjunction between cultural and social structures; one being devoted to apocalyptic attitudes, the other to technocratic decision-making. What is even more surprising is our tolerance for the contradiction between myth and social structure. I do believe that this is possible only because we have a ritual by which and with which we live—that, like any good ritual, cloaks for us, the participants, these contradictions between the myths (let me say of the French Revolution) and social reality (which moves in the direction of increasing polarization between rich and poor, increasing passivity, and increasing degradation of our milieu). This ritual, I believe, is school.

School is a ritual which initiates the creation myth of our

time: the myth of unending consumption. This modern myth is grounded in the belief that process inevitably produces something of value, and that therefore production necessarily produces demand. School teaches us that instruction produces learning. The existence of schools produces the demand for schooling, and once we have learned to need school all our activities take the shape of client relationships to specialized institutions which make us better because they give us their treatment. Once the self-taught man has been discredited, all nonprofessional, noncertified activity is rendered suspect. Hope, which Pandora kept in her box, is replaced by rising expectation. In school we are taught that valuable learning is the result of attendance, of sitting, and that this value increases with the amount of input. As a matter of fact, every child knows (not, any more, the parent who sends him to school, but the child) that learning is the human activity which least needs manipulation by others, because even now, in a highly schooled society, most valuable learning is not the result of specific instruction by certified teachers. It is, rather, the result of unhampered participation in a meaningful context. Most people learn best by being "with it," by hearing their maternal tongue spoken, but school makes man identify his personal cognitive growth with elaborate planning and manipulation. And once man has accepted the need for an institutionalized value in order to grow, he easily can be hooked on the need for any other institutional process. This transfer of responsibility, this ritual transfer of responsibility from self to institution, guarantees social regression—especially once it has been accepted as an obligation incumbent on each citizen.

I want, therefore, for a moment, to look at four aspects of the myth that schooling—the idea that I cannot be a full citizen without consuming the services of a large institution —characterizes.

First of all, the Myth of Measurement of Values. The institutionalized values school introjects are quantified values. School initiates man into a world where everything can be measured, including man's imagination and, indeed, man himself. As a matter of fact, personal growth does not seem to be a measurable entity; otherwise it wouldn't be personal, unique. It should be—I assume, at least, and here is a value from which I have to start—growth in disciplined dissidence; this growth cannot be measured against any rod nor compared as a copy to someone else's achievement. Such learning can only emulate others in imaginative endeavor. It can lead to following their footsteps but not to mimicking their gait. The learning I think we have learned to prize in Western civilization is immeasurable re-creation. But school pretends to break up learning into subject matters, to build a curriculum made out of these prefabricated blocks and put them into the pupil, and to gauge the result on an international scale. School creates sixteen classes of dropouts in the United States and in Bolivia as well, only in Bolivia everyone knows that Bolivia is much inferior in comparison to the United States, where you can't join the sanitation union in New York anymore unless you have a high-school degree. When I say this in Latin America people think that I am crazy. And especially so if I tell them that it is the union which insists on it.

A man who submits to others for the measurement of his personal growth soon applies the same ruler to himself: he no longer has to be put into place, into his place, but puts himself into his assigned slot—squeezes himself into the niche which he has been taught to seek, and the very process puts his fellow into his place, until everyone and everything fits. A man who has been schooled down to size lets unmeasured experience very easily slip out of his hands. What cannot be measured becomes second-class for him. He does not have to be robbed of his creativity; under instruction, he has unlearned to do his thing, to be himself, and values only what has been made or could be made—and that is what he envies. Once a man has the idea schooled into him that value can be produced (learning itself) and measured (by the level at which he dropped out of school), he tends to accept all kinds of rankings. Very easily, then, there is a scale for international development, another for the intelligence of babies, and even progress toward peace can be calculated in body counts. In a schooled world the road to happiness must be paved by some kind of consumer index, and school not only ritually hooks us on consumption of value, of institutionalized value, and on measurability of value, but also on packaging of value. School sells curriculum—a bundle of goods made according to the same process and having the same structure as any other engineered merchandise. Curriculum production begins with scientific research—so-called scientific research—on whose basis educational engineers predict future demands and tools for an assembly line called the school system, within the limits set by budgets and taboos. The curriculum designer, the practical educational administrator, then packages the components of educational research for local consumption, after which the principal sets up the marketing strategy for selling the goods to parents, to taxpayers, and to children. Finally, the distributor of this commodity, the teacher, delivers the finished product to the consumer, the pupil, whose reactions he carefully studies in order to provide research data for preparation of the next model, which the researcher will produce. Ungraded, student-designed, team-taught, visually-aided, issue-centered—however you call the new curriculum —the result of the curriculum production process looks like any other modern staple: it is a bundle of planned meanings, a package of values, a commodity whose balanced appeal makes it marketable to a sufficiently large number to justify the cost of production. Of course, we have now many free schools which pretend that they can produce handcrafted curricula for special circumstances at an even higher cost. This is just Parkinson's Law: you can produce education at any stated cost at which you desire to produce it, and learning difficulties necessarily rise with the money available for instruction.

The consumer-pupil is taught to conform his desires to marketed meanings and made to feel guilty if he does not behave according to the predictions of consumer research. Educators can justify more expensive curricula whenever they want. I had a very good example of this about six months ago. I was in France with a group of educators and told them that France was very quickly approaching the level of $350 per capita per student all through the first ten grades, and at that level of investment reading difficulties make a jump, that from a minority of two percent they become a problem for twenty percent of students. Well, the French colleagues to whom I spoke replied, *Monsieur, vous êtes fou! Ça peut se passer aux Etats-Unies.* This can happen in the United States, but not here in France! A month ago I was back with the same group, and very shamefacedly one gentleman from the *Ministère* took out a report indicating that the level of expenditure per student had been reached finally—and twenty-three percent of the students in the *lycées* had major reading difficulties.

School also trains us for, allows us to enact ritually, the Myth of Self-perpetuating Progress. Even when accompanied by declining results in learning, paradoxically, rising per-capita instructional costs increase the value of the pupil in his own eyes and in the eyes of others as reflected in the market. Ivar Berg has brought this to light very well in his book *The Great Training Robbery,* in which he points out that there is a very close correlation in the United States between income and status and the corresponding degree, but not between job efficiency and the degree. At almost any cost, school pushes the pupil up the levels of competitive curricula consumption into progress to ever-higher levels. Expenditures to motivate the student to stay in school become increasingly higher—they skyrocket!—as he climbs the pyramid. On lower levels they might purchase school lunches or lollypops, but on higher levels they're hidden in new football stadiums, chapels, or a program such as "international education" (high-class tourism). If it teaches nothing else, school teaches the value of escalation—the value of the American way of doing things, be it transportation (more speed into more cars and less mobility) or more expenditures on wars (rising cost of delivering the very cheap bullet in order to get a dead Vietnamese). School programs hunger for progressive intake of teaching, but even if the hunger leads to steady intake it never yields the joy of knowing something to one's satisfaction. Each subject comes packaged with the instruction to go on consuming one "offering," as they call it in Academe, after another, and last year's wrapping is always obsolete for this year's consumer. The textbook racket builds on this demand. Educational reformers promise each new generation the latest and the best. And the public is schooled into demanding what these people offer.

Both the dropout, who is forever reminded of what he missed, and the graduate, who is made to feel inferior to the new breed, know exactly where they stand in the ritual of rising deception, and are set to support a society which euphemistically calls the widening frustration gap a revolution of rising expectations. We see this very clearly in Latin America: at the end of the Alliance for Progress we have definitely overcome the major obstacle to schooling, which UNESCO so clearly defined in the early sixties as the people's unwillingness to obligate their children to go to school. Today no Latin-American country spends less than one-fifth of its budget on schooling, no country exists which hasn't made five years of school attendance obligatory, and therefore now has a right to discriminate against those who have not consumed five years of school-sitting. And with this huge amount of investment, no country in Latin America can provide school attendance up to the obligatory level for two out of three of its young people. At the same time, under U. S. technical advice from the Agency for International Development, several countries now are considering raising their minimum obligatory and free school levels—thereby disenfranchising ever larger numbers of people. The power to set minimum levels of schooling is much more crucial in keeping a modern consumer economy going than is control over the discount rate. As a matter of fact, growth perceived as open-ended consumption, as eternal progress, can never lead to maturity, and the more you obligate people to take part in the ritual of progressive and competitive consumption, which is supposed to lead to equality (obligatory competition for the purpose of equality), the more you hook them on submission in the society that is consumer-oriented.

Ivan Illich

Arnold Toynbee believes that the decadence of a great culture is usually accompanied by the rise of a new world church which extends hopes to the domestic proletariat while serving the needs of a new warrior class. I'm not going to criticize Arnold Toynbee, but start from this insight, which serves my purpose. School seems eminently suited to be the world church of our decaying culture. No ritual could better veil from its participants the deep discrepancy between social principles and social reality or structure in today's world. School is secular, scientific, and death-denying. It is of a piece with the modern world. Its classical critical veneer and the pretense of the teacher make it appear pluralist, if not antireligious. Its curriculum both defines science, and is itself defined by science, by so-called scientific research. It is a ritual example of the circularity of modern organized science. And finally, it is death-denying. No one completes school—yet. It never closes its door on anyone without first offering him one more chance all through life. Nobody can say, "Oh, I was a stupid child when I didn't finish obligatory schooling." There is always remedial, adult, or continuing education available, much more difficult to absolve than child-schooling, which reinforces the sense of guilt of those who don't have even a fifth-year certificate—as, necessarily, most of the people in the world. School serves as an effective creator and sustainer of social myth because of its structure as a graded game of ritual promotion, as a gamble. Introduction into this gambling ritual is much more important than what or how something is taught.

It is the game itself that schools. It is the game itself that gets into the blood and becomes a habit—not the quality of the teacher, of the classroom, of the selection of students. The whole society, once it accepts the rule of obligatory schooling, is initiated into the myth of unending consumption of teaching services. This happens in the measure that token participation in the open-ended ritual is made compulsory and then very soon becomes psychologically compulsive everywhere. School directs ritual rivalry into an international game which obliges competitors to blame the world's ills on those who cannot or will not play.

School is a ritual of initiation for everybody. Initiation into the Creation Myth of our society, which is that man is not hungry, doesn't have an empty stomach, but that the bottom has dropped out of his stomach, ultimately causing him to be measured as not hungry—but still he suffers from nonsatiety. School is a ritual of initiation which introduces the neophyte to the sacred race of progressive consumption. It is also a ritual of propitiation whose academic priests mediate between the faithful and the gods of privilege and power. And it is, finally, a ritual of expiation which sacrifices its dropouts, branding them as scapegoats of underdevelopment. The dropouts in Harlem or the illiterates in the world are the reason why our world is so behind what it should be, scientifically and technologically. Even those who spend at best a couple of years in school (and this is the overwhelming majority in Latin America, in Asia, and in Africa) learn to play at this strange game of culpable underconsumption. In Mexico six grades of school are legally obligatory. A child born into the lower economic third has only two chances in three to make it into the first grade, but if he makes it into the first grade, and therefore begins initiation into a sense of culpability, he has four chances out of one hundred of finishing obligatory schooling by the sixth grade. If he is born into the middle third, his chances increase to twelve out of one hundred. And, of course, Mexico is more successful than the great majority of the other Latin-American republics in providing basic education.

Everywhere, all children know that they were given a chance, albeit an unequal one, in an obligatory lottery, and the presumed equality of the international standard now compounds their original poverty with their self-inflicted discrimination, which they accept at the moment they define themselves as dropouts. From above, schooling is a selection system; from below, it is a degrading system. The dropout has been schooled to the belief in rising expectations and can now rationalize his growing frustration outside of school in terms of his rejection of the grace which school could have conferred on him. In a funny way this is a religious conception. The dropouts are excluded from heaven because, once baptized by registration, they did not go to church, to school. They were born by the decision of the modern theologian, called a pedagog, in original sin which requires treatment through the teacher. They were hooked into the first grade but now go to Hell (in the New Testament Hell and Slum are used interchangeably under the image of Gehenna), the dropout going to Slum because of his personal fault of not staying on in school.

As Max Weber traced the social effects of the belief that salvation belonged to those who accumulated wealth, so we now can observe that predestination is reserved for those who accumulate years in school. The capitalist mentality, which was accumulative in the last century, has become consumption-oriented in ours, and the ritual of it is best expressed through obligatory graded competition which promises equality. School combines the expectation of the consumer, expressed in its claims, with the belief of the producer, expressed in its ritual. It is a liturgical expression or enactment of a worldwide cargo cult, reminiscent of the cargo cults which swept Melanesia in the forties, injecting their members with the belief that if they put on a black tie over their naked torsos and hair shirts Jesus would arrive in a white steamer, bringing for each an icebox, a pair of pants, and a sewing machine. School fuses the growth in humiliating dependence on a master with the growth in a futile sense of omnipotence, so typical of the pupil who wants to go out and teach all nations to save themselves. This ritual is tailored to the stern work habits of the hard hats, and its purpose is that of celebrating the myth of an earthly paradise of never-ending consumption, especially of services, which is the only hope for the wretched and dispossessed and, a priori, impossible.

Epidemics of such insatiable this-worldly expectations have occurred all throughout history, especially among colonized and marginal groups in all cultures. Jews had their Essenes and Messianists, serfs in the Middle Ages their Thomas Muenzer, dispossessed Indians from Paraguay to Dakota their infectious dances which would lead them flying into paradise. These sects were always led by a prophet, and limited their promises to a chosen few. The school-induced expectation of the kingdom, to the contrary, is impersonal rather than prophetic and universal rather than local. Man has become the engineer of his own messiah, and promises the unlimited rewards of science to those who submit to progressive engineering of this kingdom-to-come. In other words, we can conceive of schooling as the ritual of alienation, and I think that this in its radical way will soon suddenly arise and emerge in consciousness.

School is not only the new world religion, it is also the world's fastest-growing employment market and should be seen as a reinforcing extension of the international caste system to which it belongs.

School gives unlimited opportunity for legitimated waste, so long as its destructiveness goes unrecognized and the cost of palliatives introduced into schools grows. If we add those engaged in full-time teaching to those in full-time attendance in school, we realize that the so-called superstructure has become society's major employer. This is often forgotten by the neo-Marxian analysts who say that the process of deschooling must be postponed or bracketed until other disorders, traditionally understood as more fundamental, are corrected by an economic and political revolution. Only if school is understood as an industry can the traditional revolutionary strategy be planned realistically, because what we traditionally conceive of as a superstructure, as a ritual, which we just observed takes place at considerable cost, obligates everybody to participation in school.

One of the insights which hopefully now are emerging is that each one of us is personally responsible for his or her deschooling, and that the individual has the power to become aware of the ritual as being a ritual, to develop a sense of the comic or tragic as a spectator in front of a theater in which he himself is involved. No one can be excused if he fails to liberate himself from schooling. Men could not free themselves from the crown until they had freed themselves from a church which claimed a right to establishment. They cannot free themselves from progressive consumption until they free themselves from the belief that they need obligatory schooling.

Ivan Illich

Our fundamental set of options, therefore, are clear. Either we continue to believe that learning is a product which justifies unlimited investments, or we rediscover that legislation and planning and investment, if they have any place in formal education—in ritual initiation—should be used mostly to tear down the barriers which now impede opportunities for learning, because learning is ultimately a private personal activity. That is, I suggest that we have not to see in which way we can improve the quality of schools but that we have to move away from obligatory gratuitous schooling at all levels, that we have to begin to discuss the disestablishment of schools, which is no more shocking in our time than the disestablishment of the Church was at the time of the Founding Fathers, later on the drafters of the Constitution. Thomas Kuhn, in *The Structure of Scientific Revolutions,* says that profound dissonance inevitably precedes the emergence of a new cognitive paradigm. The facts reported by those who observed free fall, by those who returned from the other side of the earth, and by those who used the new telescope, did not fit the world view which became possible after Newton. Quite suddenly a new paradigm became acceptable. The dissonance which characterizes many of the young today is not so much cognitive, or purely cognitive, as it is profoundly attitudinal—a feeling about what a tolerable society cannot be like. What is surprising about this attitudinal dissonance is the ability of a very large number of people to tolerate it. Their capacity to pursue incongruous goals must be clearly explained, and I try to do exactly this by following Max Gluckman, who believes that all societies have produced rituals to hide such dissonances from their members. Rituals can hide from their participants even those discrepancies and conflicts which are extreme and fast-growing. School, in fact, does hide from both teacher and educator the rising frustration gap which it produces, and makes them believe that they are working toward equalizing opportunities. HEW studies and other, independent studies of the University of Miami (published in 1970) show that in order to provide equal educational opportunities on the grammar and high-school levels in the United States one would have to add about eighty billion dollars to the forty billion then being spent. I suggest that we ponder this, and also the one billion dollars spent on Title I of the Poverty Act each year during a three-year period which, according to 180 different studies collated and compared at the University of Michigan, actually increased the relative educational disadvantage of the six million children for whom the money was used, and at the same time, during the same period, brought to light ten million additional children who were educationally disadvantaged. As long as individuals are not explicitly conscious of the ritual character of the process through which they are initiated into a society of competitive consumption, the forces which shape their cosmos, they cannot break the spell and shape a new cosmos—that is, as long as we are not aware of the ritual through which school shapes the progressive consumer who is the economy's major resource, we cannot break the spell of this economy and shape a new one.

I believe, therefore, that it is by now a self-fulfilling prophecy to say that the vulgar dramatization of obligatory schooling will lead to major disenchantment within the next few years. We will thereafter move from radical reformism, now so prevalent in the schools of the United States, to a movement toward the disestablishment of obligatory schooling. This, I dare to predict, will take the form of a demand for three major guarantees of civil freedoms, which will be sought in a variety of ways and which all constitute statutory implementations of freedoms guaranteed by the Constitution.

We need a guarantee against regressive taxation, which will find its expression in individual entitlements to educational resources, to public funds destined for educational purposes. It is intolerable that at the moment in the United States the richest ten percent of the population receives ten to twelve times the public funds per capita for the education of their children as are given to the poorest ten percent; practically all private funds go to this same upper ten percent minority, in addition. (Of course, this is not nearly as serious a situation as in Mexico, where every university student obtains 350 times more money for his education

Ivan Illich

than the student coming from an average family; or in Bolivia, where the disproportion is one to 1,400 and every third student in high school is the child of a teacher.)

A second radical guarantee of freedom probably will be the inclusion of a new addition to civil rights, especially antidiscrimination. At this moment I know that it sounds strange—as strange as it would have sounded just several decades ago to have spoken about the right of persons to freedom from discrimination because of their color, and just a century earlier than that, about freedom from discrimination because of one's lack of any publicly professed faith in a God. We must provide North American citizens with guarantees against discrimination because of the method by which they have acquired whatever learning they have. Inquiry into a man's learning curriculum must be as private as inquiry into his sexual preferences or religious beliefs or the title of his parents. Unless we move in this direction we cannot move toward a disestablishment of schooling.

The third guarantee toward which we must move is a guarantee against the right of professional groups to define treatment goals on which disenfranchisement or enfranchisement is dependent. Increasingly, we give professional organizations the right to define what kind of treatment a man must consume in order that he may exercise the rights of a full-fledged citizen. You know the proposal of the psychiatrist who dealt with Mr. Nixon before he became President, suggesting that all children between certain ages be examined in order to identify those who have aggressive tendencies and then to provide treatment to render them less aggressive—if necessary, obligatory and full-time. Which reminds me, of course, of the great man because of whom de Tocqueville came to the United States—Horace Mann, who proposed the idea that criminals, if locked up, would become less criminal. If they are put into cages, they would be led (it would be found) in the same direction as provided by the developed public schools.

Politically, my proposal of deschooling society and not just education probably has some merit. Because it does assume—well, I guess, provide—a rationale for very strange, otherwise incompatible bedfellows to meet.

Epilog

Fear and Self-Transformation

Michio Nagai
Symposium Chairman
Asahi Shimbun, *Tokyo*

Sociologist and educator. Educated at Kyoto and Ohio State Universities. Editorial Board of the Asahi Shimbun *since 1970. Consultant to the Carnegie Corporation. Former professor of sociology, Tokyo Institute of Technology. Extensive writings in fields of sociology and education include a study completed in 1970 of the changing role of universities, aimed at redesigning the Japanese system of higher education.*

Before I go on to give my final remarks, let me say a few words about the way I look in informal clothes. I simply did not know until I arrived in this city three days ago that formal clothes were required for the banquet. Somebody suggested that I could rent a tuxedo. But, frankly speaking, I have never worn one up to this age, and to change a habit instantly is not an easy thing for me. So it is, in a way, a cultural difference reflected in the way I look tonight. It is not at all a sign of disrespect on my part to the host and to you, ladies and gentlemen.

Now the Fourth International Symposium, *"Cultural Styles and Social Identities: Interpretations of Protest and Change,"* has come to an end. I would like to extend my gratitude to all the participants for your cooperation, which helped to make this symposium successful.

At this symposium there have been many opinions expressed. I have heard it expressed that this symposium was too broad in its concept, and too narrow. That there were too many contributors invited, and not enough. My colleagues have presented us with wise and lofty thoughts, and they have shared with us in impromptu party pleading, invective, and the give and take of separate informal meetings. We have even almost heard from the "silent majority."

We have enjoyed the Smithsonian Institution as a gracious host, and for having provided a liberal arena for this exchange and cross-fertilization of ideas. In addition, many individuals have contributed generously toward making our visit a comfortable one. Thus, speaking for my colleagues and myself, I thank you all for this opportunity.

It seems to me to have been a very successful symposium. We shall each carry away our own ideas, reactions, and impressions of it all. With this in mind I wish to share with you my own feelings and thoughts coming out of the swift hours since my arrival here.

Having listened to the lectures and discussions, and having read the papers prepared for the symposium, I have learned a great deal. Although there is little I could add to the wise words spoken by the participants, the strong impression I had while listening to these words was that seldom in the history of mankind has man been so uncertain about himself as he is today.

But I must say that this impression did not come to my mind suddenly at this symposium; it is one I have entertained for several years while teaching at universities in Japan, in the United States, and in various other parts of the world. A question in my mind, therefore, has been why it is that man is so uncertain about himself as he is today, including, of course, the youth in various parts of the world, minority groups in some highly industrialized societies, peoples of less powerful and smaller nations, and sometimes even those who are in positions of power.

To my way of thinking, such widespread and deeply rooted uncertainty and anxiety exist in the minds of millions of people on earth because, underlying all sorts of uncertainties and identity crises, there must be basic and serious problems in the present stage of the history of mankind.

Let me make my point clear by telling you first a simple and concrete fact as a person coming from Japan. Shortly before her surrender in the Second World War, nuclear bombs hit two cities of Japan. Hence, we became the first and only nation on earth to experience the power of nuclear weapons.

Let me remind you that I am pointing out this fact not with any nationalistic sentiment, but with the aim of posing a universal problem for mankind. It was, in fact, true that there was very little sentiment of revenge because of the incident. The Japanese really did become a peace-loving nation precisely because they learned by experience how terrible war can be at this stage of historical development.

The majority of Japanese were happy to see the New Constitution enacted by the advice of the Allied Forces, in which the nation of Japan declared the denunciation forever of war of any type in the famous Article 9. We then were really convinced that we would pioneer a new world by building up a state without any armament.

Twenty-five years have passed since then. The New Constitution still stands in the original form. However, nobody can deny the simple facts that Japan has gradually built up self-defense forces, that the danger of thermonuclear war is constantly with us all, and that the danger does not seem about to vanish in a foreseeable future.

One wonders, then, what it is which maintains this danger. The nuclear weapon is a child born out of the marriage of two factors; that is, the absolute sovereignty of the state and the ever-progressing processes of industrialization. If we are to survive by doing away with the dangers of war, there are at least three questions we must examine closely: (1) Is it the particular combination of absolute state sovereignty and industrialization which is wrong? (2) Is the present form of state sovereignty outdated? (3) Is there anything wrong in the basic assumptions underlying the processes, structures, and values of industrialization?

I do not have to tell you that these are basic historical questions. For roughly three centuries the modern sovereign states have been safeguarding their people. Industrialization has been going on for about two centuries, and it has been helpful in raising the living standards of people and providing comforts and conveniences.

If I may talk again of my country, after so much economic growth and technological innovation in the postwar period, however, the natural as well as the cultural environments of Japan are polluted and damaged for the first time in our 2,000-year history. Talking with students in Japan, in the United States, and in other parts of the world, I am convinced that some significant changes are indispensable in the form of state sovereignty and in the style of industrial cultures.

I am not capable of offering any persuasive answers to these basic questions. The establishment of the League of Nations and, later, of the United Nations, were attempts to modify the existing structure of state sovereignty, but we all know that the real power of major states has been only slightly checked so far. Many more creative efforts will be necessary to build up a worldwide political structure suitable for our time.

The basic assumptions concerning values, processes, and structures of industrialization seem to neglect (more than anything else) the importance of man, his cultural heritage and nature.

It will be many more years before the survival of man and nature is really assured. It is not at all strange that peoples of different countries, especially the youth, are so disturbed by the unstable and dangerous social order of today. The present situation of the world reminds me of the social disorder before the achievement of the Glorious Revolution in seventeenth-century England. The problem then was far smaller in scale, and was confined solely within one country. Thomas Hobbes called it a state of natural condition where man's survival was at stake.

With the sense of fear in his mind, Hobbes observed and analyzed the social confusion of his country and finally designed a model sovereign state which was to guarantee the safety of man.

To have fear of our time is not a sign of weakness. Fear can be turned into strength—strength on the basis of which man can calmly look at and rebuild the world around him. Above all, man's fear of himself is a part of the unstable social order—is the starting point of self-transformation.

Let us hope that there will be some among us who are not afraid of alienation from the confused social order of today, and who, with patience and with creativity, hopefully will design a new social order in which the offspring of all of us will be able to find the meaningful value of human existence.

Michio Nagai

Suggested Further Reading

(The following are but a fraction of published and unpublished literature on diversity, identity, protest, and change. For more comprehensive lists, including periodicals, consult The Center for the Study of American Pluralism, National Opinion Research Center, 6030 S. Ellis Avenue, Chicago, Illinois 60637.)

AHLBORN, RICHARD. *Penitente Moradas of Abiquiú.* Contributions from the Museum of History and Technology, paper 63. Washington: Smithsonian Institution, 1968.

ARENDT, HANNAH. *On Revolution.* New York: Viking Press, 1963.

ARENSBERG, CONRAD M., and SOLON T. KIMBALL. *Culture and Community.* New York: Harcourt, Brace & World, 1965.

BLAKE, HERMAN. "Black Nationalism," in *Protest in the 1960's, Annals.* Robert Rosenstone and J. Boskin, editors. Philadelphia: American Academy of Political and Social Science, March 1969.

BOORSTIN, DANIEL. *The Americans: The Democratic Experience.* New York: Random House, 1973.

——————. *Democracy and Its Discontents: Reflections on Everyday America.* New York: Random House, 1974.

BROMLEY, YU. V., editor. *Soviet Anthropology and Ethnology.* (Compiled from papers presented at the IXth International Congress of Anthropological and Ethnological Sciences, Chicago, 1973.) The Hague: Mouton, 1974.

CLIFFORD, GEERTZ. *The Interpretation of Cultures.* **New York: Basic Books, Inc., 1973.**

DELORIA, VINE, JR. *Custer Died for Your Sins.* New York: Macmillan, 1969.

——————. *We Talk, You Listen.* New York: Macmillan, 1970.

——————. *God is Red.* New York: Grosset & Dunlap, 1973.

EWERS, JOHN C. *The Indians of Texas in 1830.* Washington: Smithsonian Institution Press, 1969.

FRIEDAN, BETTY. *The Feminine Mystique.* New York: Norton, 1965.

GLAZER, NATHAN, and DANIEL PATRICK MOYNIHAN. *Beyond the Melting Pot.* Cambridge, Mass.: MIT Press, 1963.

GREELEY, ANDREW M. *Ethnicity in the United States: A Preliminary Reconnaissance.* New York: John Wiley & Sons, 1974.

GREER, GERMAINE. *The Female Eunuch.* New York: McGraw-Hill, 1971.

GRIDLEY, MARION E. *American Indian Woman.* New York: Hawthorne Books, 1974.

HUGHES, EVERETT C. *The Sociological Eye.* Chicago: Aldine-Atherton, 1971.

ILLICH, IVAN. *Deschooling Society.* New York: Harper & Row, 1970.

KING, MARTIN LUTHER, JR. *Where Do We Go from Here: Chaos or Community?* New York: Harper & Row, 1967.

LEVI-STRAUSS, CLAUDE. *The Raw and the Cooked.* New York: Harper & Row, 1969.

MEAD, MARGARET. *Continuities in Cultural Evolution.* New Haven: Yale University Press, 1964.

MORGAN, ROBIN, editor. *Sisterhood is Powerful.* New York: Random House, 1970.

HANDLIN, OSCAR. *Children of the Uprooted.* New York: Grosset & Dunlap, 1968.

MOSELEY, GEORGE V. *A Sino-Soviet Cultural Frontier: The Ili Kuzath Autonomous Chou.* Cambridge, Mass.: Harvard University Press, 1966.

——————. *The Consolidation of the South China Frontier.* Berkeley and Los Angeles: University of California, 1973.

NEWTON, HUEY, and HERMAN BLAKE. *Revolutionary Suicide.* New York: Harcourt Brace Jovanovich, 1973.

NOVAK, MICHAEL. *The Rise of the Unmeltable Ethnics.* New York: Macmillan, 1972.

QUIRARTE, JACINTO. *Mexican American Artists.* Austin: University of Texas Press, 1973.

RENDON, ARMANDO. *Chicano Manifesto.* New York: Macmillan, 1972.

SAMOVAR, LARRY A., and RICHARD E. PORTER, editors. *Intercultural Communication: A Reader.* Belmont, Calif.: Wadsworth Publishing Co., 1972.

SANCHEZ, RICARDO. *Conto Ignito Mi Liberacion.* El Paso, Texas: Mictla, 1972.

SCHRAG, PETER. *The Decline of the WASP.* New York: Simon & Schuster, 1972.

SHIBUTANI, T., and K. KWAN. *Ethnic Stratification: A Comparative Approach.* New York: Macmillan, 1965.

SIMMEN, EDWARD. *The Chicano: from Caricature to Self-Portrait.* New York: Mentor, 1971.

SIMPSON, G. E., and M. Y. YINGER. *Racial and Cultural Minorities.* New York: Harper & Row, 1965.

STEWART, T. DALE. "Anthropology and the Melting Pot," *Smithsonian Annual Report,* 1946; also adapted for *The Smithsonian Treasury of Science,* volume 3, Webster P. True, editor. New York: Simon & Schuster, 1966.

──────. *The People of America.* New York: Scribner's, 1973.

STURTEVANT, WILLIAM C., general editor. *Handbook of North American Indians.* (Smithsonian Institution; in preparation for 1976.)

TOURAINE, ALAIN. *The Post-Industrial Society.* New York: Random House, 1971.

USDIN, GENE, editor. *Perspectives on Violence.* New York: Brunner/Mazel, 1972.

VOGEL, VIRGIL J., editor. *This Country Was Ours: A Documentary History of the American Indian.* New York: Harper & Row, 1974.

WASHBURN, WILCOMB E. *Red Man's Land — White Man's Law.* New York: Scribner's, 1971.

──────. editor. *The American Indian and the United States: A Documentary History,* 4 volumes. New York: Random House, 1973.

ZARETSKY, IRVING, and MARK LEONE. *Religious Movements in Contemporary America.* Princeton, N.J.: Princeton University Press, 1974.